Sunk!

Sunk!

How the Great Battleships were Lost

by DAVID WOODWARD

London
GEORGE ALLEN & UNWIN
Boston   Sydney

George Allen & Unwin (Publishers) Ltd,
40 Museum Street, London WC1A 1LU, UK

George Allen & Unwin (Publishers) Ltd,
Park Lane, Hemel Hempstead, Herts HP2 4TE, UK

Allen & Unwin Inc.,
9 Winchester Terrace, Winchester, Mass 01890, USA

George Allen & Unwin Australia Pty Ltd,
8 Napier Street, North Sydney, NSW 2060, Australia

First published in 1982

British Library Cataloguing in Publication Data

Woodward, David
  Sunk!: how the great battleships were lost.
  1. Battleships—History
  I. Title
  359.3′252   V763

  ISBN 0-04-359009-8

Set in 11 on 12 point Times by Grove Graphics, Tring
and printed in Great Britain
by Mackays of Chatham

*Another One
for the
Elizabeths*

Acknowledgements

I wish to acknowledge with especial gratitude permission given me from the following sources to use vital material for inclusion in *Sunk!*

The Proceedings of the U.S. Naval Institute, dealing with the Sinking of the *Yamato*, the sinking of *Kongo* (from U.S. Submarine Operations).

John Murray, for material from *Rasplata* and *The Battle of Tsushima* both by Vladimir Semenoff.

Curts Brown, in respect of *The Fleet that Had to Die* by Richard Hough.

Seeley Service Ltd, c/o Frederick Warne, *The Devil's Device* by Edwin Grey.

Arbeitskreis für Wehrforschung, three articles from Marine Rundschau. SMS *Lützow* in der Skaggerakschlacht (MR May 1926), Versenkung des *Viribus Unitis* (August 1936), and "Wusten die Italiener . . ." (July 1979).

Penguin Books, *Battleship*, Martin Middlebrook and Patrick Mahon.

Purnell, *Ironclads in Action*, H. W. Wilson, Sampson Low.

Souvenir Press, *Black Saturday*, A. McKee (Loss of HMS *Royal Oak*).

Three Correspondents in France who have furnished me with hitherto unpublished accounts of the loss of battleships *Iéna* and *Liberté*, Commandant J. C. Bertout, Commandant Dreuzy and Monsieur Dussau of St Germain-en-Laye.

Contents

List of Illustrations

(between pages 82 and 83)

Introduction

The age of the battleship, whose most powerful weapon was the gun, lasted from the American Civil War until the last stages of World War II. During those eighty-five years between five and six hundred ships which could be more or less correctly classified as battleships, or alternatively as capital ships or battle-cruisers, were built for the following navies:

Britain	174	Spain	6
France	83	Brazil	5
Germany	77	Argentina	3
USA	57	Chile	3
Russia	46	Greece	3
Italy	46	China	2
Japan	29	Portugal	1
Austria-Hungary	10		

In addition, coast defence vessels of a smaller size carrying three or four heavy guns and able to operate in shallow waters were built for Norway, Sweden, Denmark, the Netherlands, Finland and Siam.

The numbers of these ships cannot be absolutely relied upon because no one has ever succeeded in defining 'battleship' to the complete satisfaction of all powers possessing them. For example, during the 'Disarmament Years', between 1919 and 1937, politicians and naval experts attempted in vain to settle this question at the conferences on naval disarmament held in Paris, Washington, Rome, Geneva and London.

Of the hundreds of these ships built only about a dozen survive, as museum ships or laid up in 'moth balls'. Five hundred and forty-five ships were built, of which 130 were sunk or wrecked in peace or war, including two which were sunk and raised, sunk once more and, once more, raised, while one ship, the Italian *Cavour,* was sunk and raised no less than three times.

Apart from those stricken by the passage of years or the decisions of international disarmament conferences, battleships perished by gunfire, torpedoes, mines, collisions – including ramming in action – shipwreck, storm, internal explosion – caused either by accident or sabotage – while some were scuttled by their own crews, and others were expended as targets for shells, bombs or torpedoes in experiments to test new weapons or means of protection.

1

Yamato, *the Last of the Many, 1945*

In the hundred years between 1855 and 1955 the twenty richest maritime nations had provided themselves with some 650 vessels which might fairly have been classified as battleships at the time they were built. Of these ships 130 were sunk, wrecked or irreparably damaged in war or peace, 80 of them by enemy action, 22 by accident and 22 scuttled by their own crews.

The first of the 130 ships sunk was the Italian *Re d'Italia*, rammed during the battle of Lissa on 20 July 1866 by the Austrian flagship *Erherzog Ferdinand Max*. The last to be sunk in action was the Japanese *Yamato* on 7 April 1945. This ship and her sister *Musashi* were the two largest battleships ever built. Their full load displacement was 72,000 tons, compared with the corresponding figure for the British *Vanguard* (50,000 tons), the German *Bismarck* (48,000 tons) and the ships of the American *New Jersey* class (52,000). The biggest of the American aircraft carriers – built after World War II and to date the holders of the all-time record for size of warships – are of 91,000 tons.

On 7 December 1941 the Japanese surface fleet comprised 10 battleships, 10 aircraft carriers, 6 seaplane carriers, 18 heavy cruisers, 18 light cruisers and 113 destroyers. On the morning of that day the backbone of the United States Pacific Fleet, the only rival of the Japanese for command of most of the Pacific and Indian Oceans, had been broken by the great carrier-borne raid on Pearl Harbor. All eight of the American battleships present there had been disabled, four of them sunk.

Forty-one months later, in April 1945, the Americans were landing on the island of Okinawa, only 400 miles from the home islands of Japan, following an amphibious advance of 6,000 miles, and they were now on the last lap before the invasion of Japan itself. After some of the fiercest battles in naval history there was left to the Japanese fleet, for its last fight, an effective strength in surface vessels of one battleship, one light cruiser and eight destroyers.

These ships were to challenge the victorious Americans in the waters of Okinawa itself, under conditions which left no hope of anything but death in battle. There could be no return, even in defeat, for the blockade of Japan by US submarines had been so rigorous that there was not enough fuel in the Island Empire to permit a return home after the sortie to Okinawa, even if destruction at the hands of the American fleet could have been avoided.

In the 1930s, when the *Yamato* and her four sister-ships were designed, the most important consideration in the minds of the Japanese naval planners was that, while Japan would never be able to afford to outbuild the United States numerically at sea, she could do so qualitatively, because the size of the American ships would always be limited by their need to pass through the Panama Canal; on the other hand the great size of these Japanese ships, of which only two were actually completed as battleships, enabled them to carry the most powerful armament ever seen in the history of naval ordnance. In World War I the British had developed an 18-inch gun, but only three of them were manufactured, and no ship mounted more than one. The *Yamato* and *Musashi*, both named after provinces of the Japanese Empire, each carried nine 18-inch guns, weighing 180 tons apiece and firing a shell weighing a ton and a half with a range of 22½ miles. This compares with the 16-inch guns, which were the biggest in the British and American navies and which fired a 2,000-lb. shell with a range of 17 miles.

The ships had armour belts more than sixteen inches thick and a deck of eight inches. They had double and, in places, triple bottoms which were protected by armour more than three inches thick, part of what was considered the most complete system of water-tight compartments ever designed, with 1,174 separate compartments in each ship. To hide the berths where the ships were being built vast camouflage nets were woven, each nearly two miles long and weighing 400 tons. This led to protests from the Japanese fishermen, who complained that the nets took so much sisal that they could neither repair their old nets nor obtain new ones.

Japanese security concerning the building of these ships was good. As late as the end of 1943 the Allies did not know how many battleships Japan possessed, and it was not until the war was over that they discovered the true calibre of the 18-inch guns. Such was the secrecy with which these guns were surrounded that Vice-Admiral Takeo Kurita, who led the ships which mounted them in battle, maintained after the war that he had not known their true size. And it was only after the *Yamato* had been in service for nearly twelve months that the Japanese permitted the German naval attaché, Vice-Admiral Paul Weneker, to see the ship.

The most striking feature of these vessels was their foremast, which looked like a futuristic block of flats twelve storeys high. A lift holding four men ran up through an armoured shaft within the mast which also contained the wiring needed for power and communications.

Nearly everything connected with the building of these ships brought its own special problems. Thus the transport of the guns and turrets of their main armament, weighing 2,774 tons each, from the factories where they were made to the yards where the ships themselves were built, at Kure and Nagasaki, required the construction of a specially designed fleet auxiliary, the *Kashino*, of 10,000 tons. Dry docks in which they were built were specially excavated and served by new floating cranes with a lift of 450 tons each.

In addition to the five ships of the *Yamato* class originally planned, work was begun on the design of two even more powerful ships, to mount 20-inch guns. However, after the battle of Midway in June 1942, when four of the biggest Japanese carriers were sunk by American carrier-borne aircraft, it was clear that the carrier would be the surface ship to decide the Pacific war and the Japanese ship-building effort was concentrated on carriers and escort vessels.

Besides laying down new carriers, ships already in existence or building were converted to carriers, among them battleships – including *Shinano*, the third ship of the *Yamato* class – cruisers, fleet auxiliaries, passenger liners and freighters.

In the meantime, *Yamato* and *Musashi* had been completed as battleships, *Yamato* a week after Pearl Harbor and *Musashi* in August 1942, by which time the former had been the flagship of the then Japanese C-in-C, Admiral Isoruku Yamamoto, at Midway, the first of the great carrier battles of history. In these battles the ships of neither side were in visual touch but attacked each other with bombs and torpedoes from the air. At Midway Yamamoto was obliged to listen, helplessly, as one by one his carriers were sunk, unable to intervene with signals or orders, lest he give away his own position.

It was the determination that the Americans should not be caught like this that caused the C-in-C of the US Pacific Fleet, Admiral Chester Nimitz, to control his fleet from a shore base and later Admiral Soemu Toyoda, who became C-in-C of the Japanese Combined Fleet in May 1944, followed his example and set up his headquarters near Tokyo.

After Midway and the battles around Guadalcanal in the Solomons, which lasted from August 1942 until February 1943, the Japanese remained on the defensive for the rest of the war, rebuilding as best they could their naval and air fleets, while the Americans worked their way across the Pacific.

During this time the *Yamato* and *Musashi* were both torpedoed by American submarines; the *Yamato* on Christmas Eve 1943, north-west of Palau by *Skate,* and *Musashi* off Truk by *Tunny* on 29 March 1944. Both ships were saved by their excellent anti-torpedo protection and damage control, and were repaired in time to take part in the greatest battle in the history of naval warfare, that of Leyte Gulf, fought between 23 and 26 October 1944. Leyte Gulf derives its rank as number one among naval battles from the fact that taking part there were 224 warships, with a total of 2,014,890 tons, while at Jutland, hitherto the biggest battle, there were 254 ships with a combined displacement of 1,616,836 tons. The tonnage of ships lost at Leyte was about twice that lost at Jutland.

On 17 October an advance party of American Rangers landed in Leyte Gulf, in the heart of the Philippine archipelago, and Toyoda, who had been waiting to see where the enemy's next attack would fall, decided that the time had come to commit everything that he had in a final effort to stop the Americans.

When the war was over Toyoda explained the reasons for his decision. If the Philippines were lost, he said, the fleet would be useless, for if it were based in home waters it could not be supplied with oil which could only come from the East Indies, and it would be a simple matter for the Americans in the Philippines to cut off that traffic. On the other hand, if the Japanese fleet were based in the East Indies the Americans in the Philippines would be able to stop its supplies of stores and ammunition which could only come from Japan. Accordingly, in Toyoda's own words: 'There would be no sense in saving the Imperial fleet at the expense of losing the Philippines.'

Toyoda then made his plans, which were based on the fact that the American forces in the Pacific operated in two parts. One was the battle fleet, at this time referred to as the Third Fleet, composed of large, fast carriers and fast, new battleships, commanded by Admiral William F. Halsey Jr flying his flag in the battleship *New Jersey.* The other was a support fleet (the Seventh Fleet) commanded by Vice-Admiral Thomas C. Kinkaid, in the amphibious force flagship *Wasatch*, with old, slow battleships and small escort carriers intended to protect and support amphibious assaults.

The Japanese plan was to separate the two American forces, lure the battle fleet away and then fall on the support fleet, destroy its transports and landing craft – the capital equipment of amphibious operations – and so set back by months the timetable of the Allied advance on the home islands of Japan.

By this stage in the war it was clear to the Japanese government that the only hope their country had of avoiding defeat was a growth of war weariness in the United States, and this war weariness might be developed if the American people found themselves faced with an extra year or more of war, while another vast invasion fleet was built to replace that which now lay at risk in the waters of Leyte Gulf.

The hope that American resolution for war might be so weakened that there was the possibility of a compromise peace was as old as the history of the United States itself. It had been cherished at one time or another by nearly all the opponents of the United States, from the British in the American War of Independence and the War of 1812 to the Germans in the two World Wars and the Communists of Vietnam.

The landings in Leyte Gulf began on 17 October, and General MacArthur's famous 'I have returned' followed on 20 October. The landing was on a scale comparable with the landings, four months previously, on the coast of Normandy. By the end of A+1[1] there were more than 100,000 troops ashore, with 100,000 more to follow. These troops had been put ashore and were now being supported by the 738 ships of the Seventh Fleet, of which 157 were warships, 420 landing craft, 84 patrol, minesweeping and survey craft and 73 service vessels of different sorts.

A curiosity of the American organisation of this campaign was that, although the two fleets and the Army were to co-operate in the same narrow stretch of water, there was no overall command nearer than Washington and the American Joint Chiefs of Staff. Under them came Nimitz as C-in-C Pacific and MacArthur as C-in-C South-West Pacific. Then, under Nimitz was Halsey, and under MacArthur was the Army and Kinkaid.

Toyoda's plan – the code name of which was SHO GO (Operation Conquer) – to destroy this force was based upon the American organisation and the geography of the Philippines, which consist of a chain of more than seven thousand islands running north and south for a distance of a thousand miles. Between the islands there are only two passages possible for heavy ships, namely, San Bernardino Strait between the islands of Luzon and Samar at the centre of the island chain, and further south, Surigao Strait between the islands of Leyte and Mindanao. At the time of the landing Halsey's Third Fleet was in position to guard San Bernardino and

Kinkaid's Seventh was watching the exit from Surigao. Through these gaps the Japanese planned to send three forces to attack the American beaches from north and south.

The biggest of the Japanese fleets under Vice-Admiral Kurita with his flag in *Yamato* comprised 5 battleships, 12 cruisers and 15 destroyers. This force was to approach via San Bernardino while another fleet, under Vice-Admiral Nishimura with two battleships, one cruiser and four destroyers, would reach the American beaches via Surigao and would be followed, also via Surigao, by a smaller force of three cruisers and seven destroyers under Vice-Admiral Shima.

These forces would, however, be no match for Halsey's great fleet and it was thus necessary that he should be lured away. For this purpose yet another Japanese force was formed with orders to show itself off Luzon, the northernmost of the Philippine islands. Commanded by Vice-Admiral Jisaburo Ozawa, it consisted of Japan's last four effective aircraft carriers together with two battleships, which had been fitted to carry twenty-two planes each at the expense of one-third of their main armament. These planes had never been provided – a symptom of the general Japanese shortage of aircraft, which was also indicated by Ozawa's four carriers which had 116 planes on board, merely two-thirds of their joint capacity.

Even more serious, however, than the shortage of planes was the scarcity of trained pilots whose lives had been sacrificed in the defence of the Pacific islands, and for whom there were no replacements.

It was correctly assumed by the Japanese that the appearance of their carriers would be a lure which the Americans could not resist, and that Halsey would dash off northward, leaving Kinkaid at the mercy of Kurita who, on the morning of 24 October, steamed into the Sibuyan Sea.

Ozawa's part as bait worked perfectly; Halsey did dash northward leaving the San Bernardino Straits unguarded and through the gap came Kurita with the main body of the Japanese fleet, now reduced to four battleships, with six cruisers and twelve destroyers.

It had not been easy for Kurita to get this far. After leaving his base at Brunei Bay he had run into an ambush set by two American submarines, which had sunk two of his heavy cruisers, one of them his flagship *Atago*, the other being her sister-ship *Maya*, while a third, *Takao*, had been so badly damaged that she had had to return to port. Kurita then shifted his flag to the *Yamato*, which now had on board no fewer than four admirals, Kurita, with his Chief of Staff, Rear-Admiral Koyanagi, Vice-Admiral Ugaki, commanding Battleship Division One, and Rear-Admiral Moroshita, captain of

Yamato. Each Japanese battleship at Leyte was individually commanded by a Rear-Admiral.

All was now before Kurita's fleet, with the prospect of a victory which could alter the outcome of the war, but they entered the Sibuyan Sea under a series of attacks from American carrier-borne planes, flying from the Pacific across the islands – a distance of about 300 miles.

In a long, hard day's work *Yamato*'s sister-ship, the *Musashi*, had been sunk with the loss of about 1,000 men after she had been hit by ten bombs and six torpedoes, despite an anti-aircraft armament of twenty-four 4.7-inch and 130 25-mm guns. In addition, the heavy cruiser *Haguro* was damaged and obliged to return to port. The *Yamato* had been hit on the forecastle and was on fire forward, while near misses had blown a large hole in her bows, but her fighting value, apart from this, was intact.

Once through the San Bernardino Strait, according to the Japanese plan, Kurita should have met the ships of Nishimura and Shima, but there was no sign of either Japanese force, and bits of intercepted signals picked up during the night made it fairly clear that Nishimura's force had been knocked out, which was, in fact, the case. Forcing its way up Surigao Strait Nishimura's two battleships, *Fuso* and *Yamashiro*, a cruiser and four destroyers had had to run the gauntlet, in succession, of Kinkaid's 39 motor torpedo-boats, 20 destroyers, 8 cruisers and finally, the 6 battleships, five of them survivors of Pearl Harbor.

The MTBs had spent most of their war patrolling and there had not been enough time for training in torpedo work. Accordingly only one secured a hit, with a torpedo which put the light cruiser *Abukuma* out of action. In the closing stages of the battle she was finished off by the bombers of the US Army Air Force.

With the destroyers it was different. Receiving the order: 'Boil up! Make smoke! Let me know when you have fired!' they went into action by divisions of three and four boats. With a splash and a bubbling, the torpedo tracks started off on their eight minute journey to the enemy, while the destroyers turned away behind their white, tumultuous wakes to get as far away as possible from the enemy.

The *Fuso* and *Yamashiro* were both sunk, the *Fuso* blowing up with an explosion so great that she split in two, both halves of the ship blazing fiercely, drifting down the Strait until overcome by fire and water. Withdrawing from this disaster area as quickly as possible, Nishimura's cruiser *Mogami* actually collided with one of Shima's destroyers, arriving in theory in support.

Discouraged, Shima turned back while Kurita held on towards the huge concentration of American shipping in Leyte Gulf until, shortly after dawn, he met a group of six American escort carriers which, together with three destroyers and four destroyer escorts, were part of Kinkaid's covering force for the protection of the great landings.

Now the Japanese plan began to fall apart.

Instead of keeping his forces together and heading at full speed for the Gulf, Kurita signalled for a kind of 'General Chase' and his ships spread themselves out in pursuit of the Americans, who defended themselves with great spirit. The carriers had been built as merchant ships and their sides were so thin that the heavy Japanese shells frequently went right through them without exploding as they came under fire from the 18-inch, 16-inch and 14-inch guns of the Japanese battleships, as well as the 8-inch and 6-inch guns of the cruisers and the secondary armament of the battleships. To defend themselves the carriers had only one 5-inch gun each and their top speed of eighteen knots gave them no chance of saving themselves by flight. Moreover, in order to get ships, planes and pilots into service as quickly as possible they had spent only a short time working up. To avoid unnecessary risks pilots were under orders not to take off or land by night, although in fact they frequently did so. A further handicap was that, because they were intended in this operation to support troops ashore, they carried for the most part light fragmentation bombs rather than the heavy bombs and torpedoes required for use against ships. They were wearing ships to serve in at the best of times, without any refinements such as air conditioning for service in the tropics, so that off the Philippines temperatures below decks were in the high nineties, and sometimes above one hundred. They now went into action as best they could, though lack of practice and the limits of their small flight decks involved both delay and danger.

The first of the carriers to become a target of the Japanese was the *White Plains*, whose crew was fascinated to find their ship being straddled by huge yellow, red, green and blue splashes, some of them nearly 200 feet high, from the dye-marked, heavy calibre shells.

'They're firing on us in technicolour,' remarked one seaman. In addition to the technicolour hues, black and white and grey were added to the colour scheme, from the luxuriant smoke screens drawn speedily and efficiently across the blue sea and sky by men who knew what they had to do to save their ships, while with unsuitable planes, the wrong missiles and unsupported by heavy ships the Americans went into the attack, continuing to make dummy runs after they had dropped their bombs or torpedoes, thus by bluff forcing the enemy to alter course away from the carriers. However, it was impossible to avoid all losses. One carrier, *Gambier Bay*, was sunk, together with two destroyers and one destroyer escort. During this phase of the action the Japanese spread themselves out so far in pursuit of the enemy that, after two hours of chase, it took another two hours for Kurita to have them in hand once more. By 11 am he was almost completely unmanned by indecision as to what he should do next.

He changed his mind about continuing to Leyte Gulf because he thought it likely that the American transport fleet would have withdrawn. He then considered going northward to join Ozawa or, if he could not find them, he might meet Halsey. Finally he decided to return to Brunei Bay through the San Bernardino Strait, leaving Ozawa's carriers as well as a number of cruisers and destroyers to be sunk by Halsey's aircraft.

On her way back to Brunei *Yamato* was hit by two more bombs, and that was the end of the battle.

Summing up Kurita's performance in the battle, the American historian, Professor C. Vann Woodward, has written: 'What was needed on the flag bridge of the *Yamato* on the morning of the 25th was not a Hamlet but a Hotspur – a Japanese Halsey instead of a Kurita.'[2]

The total losses in this battle were:

American: 1 light carrier (*Princeton*)
        2 escort carriers (*Gambier Bay, St Lô*)
        2 destroyers (*Johnston, Hoel*)
        1 destroyer escort (*Samuel B. Roberts*)
        1 submarine (*Darter*)

Japanese: 3 battleships (*Musashi, Yamashiro, Fuso*)
        1 fleet carrier (*Zuikaku*)
        3 light carriers (*Chitose, Chiyoda, Zuiho*)
        6 heavy cruisers (*Atago, Maya, Chokai, Suzuya, Chikuma, Mogami*)
        4 light cruisers (*Abukuma, Tama, Noshiro, Kinu*)
        11 destroyers (*Wakaba, Yamagumo, Michisio, Asagumo, Hatsusuki, Akitsuki, Nowake, Hyashimo, Uranami, Fujinami, Shiranhui*)

When the war was over, Ozawa said of Leyte, 'After this battle there was no further use assigned to surface vessels, with the exception of some special ships.'

As we shall see, this did not apply to *Yamato*, but in the meantime most of the Japanese survivors of Leyte made their way back to Japan, except for the battleship *Kongo* which, with another battleship, two cruisers and three destroyers, was spotted and attacked by the American submarine *Sealion* early in the morning of 21 November. The very small destroyer escort provided for the big ships on this occasion is an indication of how the Japanese destroyer force, vital for anti-submarine protection, had been depleted by losses.

Sealion was running on the surface. Summoned from his bunk, her captain, Commander E. T. Reich, wearing pyjamas of robin's egg blue, climbed up to his bridge and carried out an immediate attack, first with six torpedoes from his bow tubes and then, from his stern tubes, fired four more. *Urakaze*, one of the escorting destroyers, was sunk and *Kongo* was hit but steamed on at twelve knots. Reich set off in chase as a gale was blowing up. The idea of a submarine chasing two battleships, two cruisers and two destroyers on the surface in the teeth of a gale would seem to be out of the question if it had not happened. The speed needed by *Sealion* to keep up and then get ahead of the enemy put too great a strain on her electric drive diesels so that, in the words of her log, she 'slowed down to full speed'. The chase went on for three hours, through the storm.

Reich took advantage of this delay to go below and change into the khaki uniform which is generally the routine wear of American naval officers. Although submariners are generally considered unconventional people apparently the captain of *Sealion* considered it not correct to sink an enemy battleship while wearing pyjamas. Then, he said, later:

> We were standing on the bridge on this wild, black night, riding the heavy seas, when, without warning, there was this brilliant flash. It lit up the entire sea for miles around, like a sunset at midnight. Just as suddenly the ship sank, and there was total blackness again.
>
> There was a horrendous explosion. A flash of light came down *Sealion*'s conning tower hatch and it illuminated the compartment below. Next came a concussion wave that plucked at breath and clothing like a vacuum cup. The boat shuddered as if shaken by a giant hand.[3]

Reich tried to continue the chase of the other ships but unhampered by the presence of the damaged *Kongo* they drew ahead into the night.

After establishing themselves on Leyte the Americans moved on to Luzon, the largest of the Philippines and situated at the head of the long chain of islands.

Japan was now to be attacked from the air, by land-based and ship-borne planes. At this moment the biggest obstacle to the long range planes of the USAAF 20th Air Force was the island of Iwo Jima, with its radar posts and its fighter bases. It was a tempting target for the next American leap forward, and in addition capture of the island which began on 19 February was long and bitter; the Japanese garrison of 22,000 men was very well dug into caves and fortifications, against them were three divisions of US Marines. The Japanese fought almost to the last man; only 212 survived to surrender, while the Marines lost 6,891 killed and 18,070 wounded. The island proved a priceless asset to the USAAF, no fewer than 2,251 planes making emergency landings upon it.

The next American move was to the island of Okinawa, only 400 miles from the nearest of the main Japanese islands Kiushu, and this saw Japanese surface ships in action for the last time in the war. The Americans, with 1,300 ships, great and small, carrying 60,000 men, advanced on the Japanese beaches in a column eight miles wide. The landings began on Easter Sunday, 1 April, and on 6 April the Japanese position on Okinawa was so grave they decided that they must commit not only their kamikaze planes but also the last of their naval strength – the *Yamato*, the light cruiser *Yahagi* and eight destroyers.

There was, at that time, available in Japan only enough oil fuel to permit *Yamato* to reach Okinawa, there being no question of her returning from this operation. The ships taking part were all partly de-stored and stores not needed for the one-way voyage were landed for the civilian population which was suffering greatly from the Allied blockade.

The plan of the operation was that the *Yamato* should run herself aground on Okinawa and use her 18-inch guns in support of the garrison of the island. On board the *Yahagi* the crew began to sharpen their bayonets for use in fighting ashore.

These ships were known as the 'Special Service Attack Force' and were under the command of Vice-Admiral Seiichi Ito. Almost incredibly they were given no air cover, the theory being that the sight of these unprotected warships would prove an irresistible lure to the American aircraft which would be diverted from their defence of the American forces coming ashore, and thus give the kamikaze planes a chance to make an unopposed onslaught.

This was the same kind of over-complicated reasoning which had resulted in Kurita being left without air cover while the scanty resources of Japanese naval air were offered, and accepted, as bait for the American carriers, with the result that four carriers and their planes were sunk. Had the *Yamato* been given air cover she might well have been able to reach Okinawa and her huge guns, heavier and of much longer range than anything the Americans possessed, might have been of decisive importance in the battle for the island which at the time was by no means as one-sided as it appears now.

The escort carrier *St Lô* at Leyte was the first ship sunk by kamikaze attack, the deliberate crashing of a fighter carrying a single bomb on an enemy ship by a pilot fully aware that he had no hope of survival. The origin of this form of attack lay in the fact that Japanese pilots realised that they were insufficiently trained to be able to drop bombs in the ordinary way with any hope of success. Altogether, according to Japanese records, 2,393 kamikaze planes and their pilots were thus expended. At the beginning of their campaign the kamikaze pilots were genuine volunteers, but later it would seem that considerable pressure was put upon them to 'volunteer'. Among these men was presumably the pilot who, on taking off on what was literally a suicide mission, shot up his commanding officer's headquarters on his way to death.

In all fifteen Allied warships were sunk and some 200 more or less badly damaged by kamikaze attacks which, in the first days of the American landings on Okinawa, assumed massive proportions with an attack by 355 planes on 6 April. A mass attack on this scale could not be repeated but individual planes continued every day to put a great strain on the fleet off Okinawa and, at one time, raised the question of calling off the operation.

On 5 April *Yamato* and her escort began to get ready for what was to be their last voyage. A stream of orders emerged from the loudspeakers throughout the ship.

'Prepare for sea.' 0815.
'All divisions bring inflammable gear topside.'
'Stow all personal belongings. Secure all watertight compart-
ments.'
'Last boat to shore.' 0831.
'All Division officers check watertight integrity.'
'Weigh.' 1000.

Thin clouds hung over the misty sea. *Yamato,* painted silver-grey, silently left Tokuyama. Officers and men alike were wearing the 'fighting khaki' uniform.

That night, according to the ship's junior radar officer, Ensign Mitsuru Yoshida, 'drinking jamborees were held in solemn gaiety', until the ship's executive officer, Commander Nomura, said: 'It has been good to see you all enjoying this day. Now our time is up.' It was a beautiful moonlit night. Last letters were written, gifts from the canteen were distributed, including cigarettes from His Imperial

Majesty which were received by officers on watch, with swords at their sides, who bowed in silence, one by one. *Kimigayao*, the Japanese national anthem, was sung, and three cheers given.

A messenger arrived in the radar compartment. 'Ensign Yoshida. Tonight's refreshment will be *shiruko*.' This was a red bean soup sweetened with sugar to which rice cakes were added, explained Yoshida in the narrative he published after the war, and he added: 'The messenger's broad smile showed his pleasure at being able to spread this news.'

After dawn there were lowering clouds, followed by a rain squall. At noon the Japanese ships were half way to Okinawa. Yoshida commented on the rations again: 'The noonday meal was simple and miserable in every respect, yet the polished white rice tasted good. It was followed by hot black tea which we drank until our stomachs bulged.'

Twenty minutes later *Yamato*'s radar picked up aircraft and from the radar room could be heard reports of the distance and bearing of approaching planes. To Yoshida it was like routine drill. The number of planes in sight rapidly increased: '2 – 5 – 10 plus, 30 plus.' Then:

A great formation roared out of the clouds and circled widely in a clockwise direction. Over one hundred enemy planes heading for us. We opened fire with twenty-four 4.7-inch anti-aircraft guns and 150 machine-guns as well as with the main batteries of the escorting destroyers. One of these, *Hamakaze*, was hit, sinking with her stern high in the air. In thirty seconds she was gone, leaving a circle of swirling white foam . . . Silvery streaks of torpedoes could be seen silently converging on us . . . From time to time *Yamato* turned parallel to their tracks.

She was now steaming at her full speed of twenty-six knots, vibrating as though her plates were going to open, listing under her helm as she zigzagged through the water, amid the noise of guns, machine-guns and bursting bombs.

The first attack began at 1220 and at 1245 the *Yamato* was hit for the first time by a torpedo forward on the port side and two bombs aft. A second attack was responsible for two more torpedo hits to port. The third attack, again by a mixture of torpedo planes and dive bombers, gave four or five torpedo hits. Everything followed so quickly and in such a confusion of noise that it was not possible for the *Yamato*'s people to make up their minds as to what was happening. Throughout the action it was clear that *Yamato*'s anti-aircraft gunnery was very poor, a fact which was afterwards attributed to the lack of ammunition for adequate target practice.

Moreover, the first three bombs had knocked out all the anti-aircraft guns abaft the beam on the port side.

At about this time Yoshida heard the Chief of Staff say, 'Judging by their skill and bravery these must be the enemy's finest pilots.' By the end of the action the upper deck was torn and holed, with scraps of metal and wrecked anti-aircraft guns scattered everywhere. After five more hits on the port side, between 1337 and 1344, Admiral Ariga, *Yamato*'s captain, ordered counter-flooding in the starboard boiler and engine rooms, to bring the ship back onto an even keel.

Yoshida said, 'Water from both the torpedo hits and the flooding valves rushed into these compartments, and snuffed out the lives of the men at their posts, several hundred in all, caught between cold sea water and steam and boiling water from the damaged boilers they simply melted away.'

Counter-flooding failed to correct the list and soon it was so great that the ship's huge battle ensign was nearly trailing in the water. 'At this moment the enemy came plunging through the clouds,' says Yoshida, 'to deliver the coup de grâce . . . I could hear the captain shouting, "Hold on men! Hold on men!" Men were jumbled together in disorder on the deck but a group of staff officers squirmed out of the pile and crawled over to the Commander-in-Chief for a final conference.'

Then Admiral Ito shook hands with the members of his staff and went below. He was never seen again. Somebody started a search for the Emperor's portrait. The sentry on the flag still remained at his post, clinging to whatever he could for a handhold. The navigator and his assistant began tying themselves to the binnacle, saying that they did not want to take the risk of floating clear and possibly surviving the ship. Others tried to do the same thing, but the Chief of Staff stopped them, hitting some who did not obey. The ship had now ceased fire, and Commander Nomura distributed sweets, biscuits and 'His Majesty's Brand' cigarettes and ordered the men to urinate.

The ship appeared to be sliding into a whirlpool 150 feet deep, and then as she finally sank there were huge flames and dark clouds seen as far as Kagoshima in Kiushiu, 200 miles away.

Altogether 386 American carrier-borne planes had taken part in these attacks. Ten of them had been lost, together with twelve aircrew. The Japanese, in addition to *Yamato* and *Yahagi*, had lost four destroyers and 3,663 men.

2

The Ram, Lissa, 1866

At the end of the American Civil War the iron battleship had arrived and it was clear that it would prove to be, for years to come, the keystone of naval power. What was not at all clear, however, was how these ships would be used and what would be the outcome of a struggle between fleets of them.

Later on the same state of confusion, ignorance and speculation concerning the future of naval warfare would exist on the eve of the First and Second World Wars and, also, at the beginning of the nineteen-eighties. Guns, mines, submarines, torpedoes, aircraft, atomic and bacteriological and chemical weapons, one by one, added to the cloud of unknowing which has existed concerning the future of naval warfare from the Crimea to the present day.

The first of these interconnecting series of puzzles was set by the events of the battle of Lissa in 1866. This was an action in which the background to the battle and the course taken by events were of a simplicity resembling that of an elementary piece of wargaming, but the incorrect conclusions drawn influenced naval thinking for nearly forty years.

The battle was fought off the island of Lissa (now called Vis) in the Adriatic on 20 July 1866. Its circumstances were curious, for it came at the very end of the Seven Weeks War, the result of which had already been decided. The war had begun on 14 June and by nightfall on 3 July it was virtually over, after the Prussians had defeated the Austrians at Sadowa or Königgratz. In the meantime, on 24 June, the Austrians had already won the battle of Custozza against the Italians but on the day that Sadowa was fought and lost the Austrians agreed to surrender Venice and the surrounding province of Venetia. Nevertheless the war did not end at once. Although the Italians, thanks to the Prussians, had won the prize of Venice there were two more pieces of territory populated by Italians and ruled by Austria which the Italians wished to liberate. One was the port of Trieste and the other the South Tyrol, and it was to get this latter area that the Italians were resolved to fight one more battle. Accordingly the government in Florence, then the capital of Italy, ordered an attack on the Austrian island of Lissa in mid-Adriatic, with the idea that, if they succeeded in taking the place, it could be exchanged at the peace conference for the South Tyrol.

Further Italian plans for the war included a landing by Garibaldi at the head of a volunteer force to stir up Croatian and Hungarian malcontents, who were then on the verge of revolt against the Austrian government.

But it was one thing for the Italian government to order the fleet to attack Lissa and another to make it do so, for their admiral, Carlo Count Pellion di Persano, was indefatigable in his search for excuses. Persano, as a junior officer, had done well against the North African pirates, and during the wars with Austria in 1859–61 he also did well, but ashore he was a boastful and bad-tempered man, fighting five duels and, as he rose in the service, he lost his daring and his willingness to take risks, although his partisans have always denied charges of cowardice.

Tegetthoff, the Austrian admiral, had achieved flag rank at thirty-nine, the same age as Nelson, in 1866. He seems first to have been marked out as a 'highflyer' when he was selected to reconnoitre, secretly, the ports at the southern entrance to the Red Sea during the winter of 1856–7. In the company of the Austrian Vice-Consul at Suez, Dr Heuglin, the two men, disguised as English tourists, made their way up the Nile to Kennar by dahabieh, then, collecting a camel caravan, they crossed the desert to Koser on the Red Sea and boarded a sombuk, a native boat so small that there was hardly room to sit upright below deck. They went on to Massowa, which the British East India Company was then suspected of wishing to buy, surveying the coast as they went. They became involved with a woman sheikh and were then mobbed by tribesmen who wounded Heuglin, held the two men virtually prisoner, robbing them of money and provisions, and leaving them just enough to get to Aden.

This strange trip had been ordered by the Austrian government because of the projected opening of the Suez Canal, planned for twelve years hence. The canal was to be of the greatest importance to Austria, for until its opening in 1869 Austrian trade with India and the Far East had been greatly handicapped by the fact that ships to and from Austrian ports had to pass through the Straits of Gibraltar and round the Cape of Good Hope to reach the Indian Ocean. However, once the canal was open there would be a direct route down the Red Sea, and somewhere on this route the Austrians wished to obtain a port. Eventually, when Tegetthoff finished his survey, it was concluded that the island of Socotra, not then in the possession of any European power, would best serve the purpose, and its price was set at 100,000 thalers or £15,000; but nothing, in fact, ever came of the expedition, save that Tegetthoff made his reputation.

In 1862 he was commanding the Austrian component of what would now be called a peace-keeping force, supervising the election of the future King George I of Greece. An indication of the way in which this election was conducted is provided by events in the constituency of Sparta, which had between six and seven thousand voters. However, when the count was taken it was discovered that the ballot boxes contained 36,000 votes – the boxes having been stolen, stuffed with forged ballot papers and then given to the returning officers for counting.

The Austrian Navy also intervened to maintain peace in the port of Smyrna, now Izmir. This operation was successful and Tegetthoff received the thanks of the city fathers of Smyrna and a present of three virgins.

In the war of 1864 between Prussia and Austria on one side and Denmark on the other, fought for the possession of the province of Schleswig-Holstein, Tegetthoff commanded the small Austro-Prussian force of two Austrian frigates and three small Prussian gunboats, which attacked and drove off a Danish squadron of three ships blockading the mouths of the Elbe and the Weser.

Persano and Tegetthoff were both heirs to political troubles when they took over their respective commands. The Kingdom of Italy and, therefore, the Italian navy, was only five years old and was formed of ships which, for the most part, had previously been part of the navies of Naples and Sardinia. There was also a so-called Sicilian fleet which had been formed by Garibaldi for his campaign in 1860, and a Venetian force made up of ships and men of Venetian origin who had deserted from the Austrian service. The merger of these forces, together with a group of officers who had formerly served in the navy of Venice, was the task of Persano and it earned for him the enmity of many officers and the ingratitude of others, a state of affairs which was to make very difficult his attempts to create a unified service.

At the same time Tegetthoff was dealing with a somewhat similar situation. The Italian troubles of 1848–9 had disrupted the Austrian navy, of which it was said at the time by the Austrians: 'We have a navy, but it's not Austrian, it's Italian.' Many of the officers were, in fact, Venetian and the cadres of the lower deck were also Venetian. The college for naval officers, where Tegetthoff and the rest of his generation were educated, was in Venice and the language of instruction was Italian.

After the troubles of 1848–9, in an attempt to deal with this,

Scandinavian officers were appointed and cadets were chosen from German Austria. This was only a stop-gap measure which more or less served its purpose, but there was a more fundamental problem to be solved, caused by the attitude of the Austrian army, which persisted in regarding the navy as an auxiliary arm of its own service. The first move in a campaign for independence was made when naval officers insisted on redesigning their uniforms so that they no longer looked like soldiers but assumed something of a naval appearance. Furthermore, a fashion grew up amongst the naval officers of doing everything possible to look unmilitary. It became their habit to walk about when ashore with their jackets un-buttoned, their hands in their trouser pockets, dragging their swords behind them, while they adopted their own style whiskers, favour-ing shaven chins and long side-whiskers known in England in those days as 'Piccadilly weepers' or 'Dundrearies'. At sea the wearing of a Turkish fez and carpet slippers and the smoking of a tschibuk passed without comment.

The Seven Weeks War was within a few days of its end before the Italian government succeeded in getting Persano to sea and start his delayed attack on Lissa, with a fleet of 12 armoured ships, 12 wooden screw frigates, 25 paddle steamers and 8 sailing ships, with a total tonnage of 122,820 tons and mounting 1,051 guns. The Austrian fleet was rather less than half this size, totalling 57,344 tons with 532 guns, mounted in 7 armoured ships, 7 wooden ships and 6 dispatch vessels and screw schooners.

Tegetthoff's best ships were two battleships of 5,100 tons each, brand new at the time of the battle, but without the main armament of modern breech-loading guns they had been designed to carry but which, having been ordered from Krupp, were seized by Prussia at the outbreak of war and replaced by obsolete muzzle-loaders.

Of the Italians, the most noteworthy were the *Re d'Italia* and *Re de Portogallo,* armoured frigates with hulls of green timber, of 5,700 tons, built in New York during the American Civil War and, greatly to the surprise of the Italians, never requisitioned by the Federal authorities.

A third Italian ship, built in England and designed to incorporate the lessons of the naval fighting in the American Civil War, was the *Affondatore.* Described as an armoured ram she was low in the water, with two pole masts and two funnels. She mounted two 10-inch guns but her principal weapon was considered to be her specially strengthened ram bow.

Waiting for the outbreak of war, Persano told his government that he would not engage the enemy until the *Affondatore* had joined his flag. This was one of his many excuses for delay, but eventually the ship reached the Italian base at Ancona and finally on 16 July Persano sailed for Lissa with his fleet and a convoy of transports carrying a landing party about 2,500 strong, intended to deal with the garrison of Lissa. This numbered 1,833, about one half of whom were gunners in the forts dominating the island. These forts had, for the most part, English names – St George, Wellington, Bentinck and Robertson, having been built and named during the British occupation of the island which had lasted from 1811 to 1814, during the blockade of the French forces in Dalmatia. Apart from the forts and their names there is, to this day, one other reminder of the British presence. Vis is the only part of Yugoslavia in which the inhabitants do not wear characteristic local costume, because, during the British occupation, communication with the mainland was impossible and the only clothes available were those imported from England.

The Italian bombardment finally began on the morning of 18 July, when the enemy was sighted by the semaphore station on Mount Hum, the highest point of Lissa, which fired three blank rounds. The Italians had preceded their attack by cutting the cable which linked Lissa with Zara (now Zadar) but the telegraphist on the neighbouring island of Lesina saw from a peak on his own island what was happening and got off a signal which was passed to Zara and from Zara to Tegetthoff who, with his fleet, lay at Fasana, near Pola (now Pula). The Austrians raised steam at once and prepared to sail. Meanwhile the Italian bombardment was well under way and the magazine of the Austrian Schmid battery blew up with an explosion which shook the whole island.

Neither side was using smokeless powder and a vast smoke cloud covered the island, lit from within by the flashes of the roaring guns. A group of Italian ships entered the harbour of St George on which stands the town of Lissa itself, but were soon driven out again by the Austrian gunners. Through the smoke it could be seen intermittently that some of the Austrian guns were out of action, that some of the earthworks had been broken up and that Fort St George was out of action.

There was a big breach in the walls of Fort Wellington, but its guns kept on firing until, at about seven o'clock in the evening, Persano drew off, leaving the Austrians to work through the night by the light of blazing torches, repairing their defences.

Next morning, 19 July, the Italians resumed their attack and Port St George was again out of action. Once more Italian ships entered the harbour and anchored within 400 yards of the shore. Their armour stood up well to the Austrian shells, but projectiles entered through open gunports and one by one the three Italian ships slipped their cables and left Port St George to the strains of the Austrian national anthem played by the Marine band of the defenders.

On this day nothing further happened for the surf was judged too strong for landing, but an attempt was fixed for the morning of the next day, which dawned gusty and rainy. Then, at about 7 am, there came out of a squall the Italian scout *Esploatore* flying the signal: 'Suspicious ships to north-west' and Persano turned to meet another foe.

It was Tegetthoff. He had sailed with his fleet at one o'clock the preceding afternoon. After the ships had cleared the entrance to Fasana opposite the island of Brioni and were in a long, single line, Tegetthoff in his flagship *Erzherzog Ferdinand Max* took up a position at the head of the line and watched the ships steam past. Without orders the men in each ship appeared on deck and in the rigging, cheering and cheering, while the bands played the national anthem and Tegetthoff waved his cap again and again in reply to each ship.

Next morning, 20 July, the day of battle, there was more mist, more rain and spray, until about ten o'clock when the clouds cleared and the Austrians saw Lissa and the Italian fleet which, as ships withdrew from their bombardment stations, formed itself into a line of fourteen grey-painted ships each with funnels a different colour for identification purposes; the whole line, over two miles long, barring the way into the harbour of Lissa, for which Tegetthoff was now heading. It was formed in three V-shaped formations each of seven ships, with the *Ferdinand Max* at the head of the first followed by seven more ships, all wooden screw, line-of-battle ships, their sides protected by anchor cables and railway rails, with the *Kaiser* in the lead, and finally there were seven little ships astern. In contrast to the Italian ships all the Austrians were painted black, and were making noticeably less smoke than the Italians, Tegetthoff having had the foresight to accumulate some 50,000 tons of the best Welsh steam coal, while the Italians had to make do with coal which was inferior and contributed more than was necessary to the dense patches of smoke from guns and funnels which lay over the sea in great clouds. Because of the smoke the *Ferdinand Max* was conned by Tegetthoff's flag captain, Sterneck, from halfway up the weather rigging, as Farragut had controlled his fleet at the battle of Mobile Bay from the shrouds of USS *Hartford*.

At this stage, as the Austrians were on the point of breaking the Italian line, Persano, for reasons which he was never able to explain satisfactorily, stopped the *Re d'Italia* in which his flag was flying and shifted to the *Affondatore*. He omitted to inform the fleet of this

move, with the result that his captains spent the entire action looking to the *Re d'Italia* for orders which never came. This was not all
the harm he did by his ill-considered action, for while the *Re d'Italia*
was stopped to allow him to transfer to the *Affondatore,* Sterneck
came out of the smoke. He saw the *Re d'Italia* before him, rang
down for full speed and headed for the enemy ship, which was in a
state of great confusion as her captain, Faa di Bruno, believing that
the *Ferdinand Max* was about to board, ordered his crew on deck to
repel attackers. Thus the Italians' main armament was left unmanned.

Sterneck ordered all his men below so that there was no one on
deck but himself and Tegetthoff as they closed with the enemy. He
altered his helm and the *Ferdinand Max* struck the Italian ship
amidships and at right angles. The Austrian's ram penetrated six
feet, smashing through armour and wooden backing alike. She then
went astern and the sea poured into a gap of fifteen square yards,
half of it below the waterline. Under the shock of the collision the
Re d'Italia had listed to starboard and then, as the *Ferdinand Max*
withdrew, rolled right over to port, capsizing in a few minutes.

The poor state of mind among the Italian crews is further illus-
trated by the fact that, as the Austrians fought their way through
the Italian fleet, on at least one occasion an Austrian ship received
the full blast of an Italian broadside through her open gun ports
without suffering any damage, because, although the Italian gun
crews had loaded their guns with the required cartridges, they had
omitted in the excitement of the battle to load shells as well.

The *Kaiser*, a wooden ship which, apart from a small, inconspicuous funnel, looked like one of Nelson's two-decker ships of
the line, was leading the second Austrian V-formation through the
Italian line when she endeavoured to follow Tegetthoff and rammed
the other Italian ironclad battleship, the *Re di Portogallo,* with the
result that she lost bowsprit and figurehead, as well as her foremast, main and mizzen topmasts, and caught fire.

In the melée now in progress the *Maria Pia,* working up to her full
speed of about ten knots, headed for the *Prinz Eugen* which turned,
so that the two ships passed each other at a distance of a few feet.
The Austrian captain Barry saw the Italian captain del Carretto and
drew his revolver. Then he thought better of it, raised his empty
right hand in salute to his Italian foe, who replied similarly.

Meanwhile another Italian ship, the *Palestro* of 2,600 tons and
classed as an 'armoured gunboat', was on fire partly, at least, because her captain had taken coal on board to increase her radius of
action. There was no room in the ship's bunkers and so the coal was
left on deck.

Sterneck attempted a second ramming but the *Palestro* avoided him, but so narrowly that a rating in the Austrian flagship seized the Italian flag and tore it down. A little later the fire in the *Palestro* reached the magazine and she blew up.

The battle ended with the entry of the Austrian fleet into Lissa harbour and the withdrawal of the Italians. Lissa was saved and the war ended but not before the *Affondatore* had also been sunk in a squall which blew up and flooded the forepart of the ship, already damaged in the battle.

The two admirals went on to their respective fates. Persano was tried by the Italian Senate, sitting as the High Court and, having been found guilty, was disgraced. Tegetthoff died of dysentery in 1871, being then forty-four years old. He was succeeded at the head of the navy by a soldier, Archduke Leopold, who boasted that he had lost 4,000 men when he was defeated at Nachod in the Seven Weeks War, and refused to take Lissa seriously, where the Austrians killed had numbered only thirty-eight.

Under such leadership or rather lack of it, the navy could not prosper and it was not until halfway through the first decade of the twentieth century that the Austrian navy once again became a force to be reckoned with. This change was brought about largely by another Archduke, Franz Ferdinand, who was to die at Sarajevo.

3

Mr Whitehead's Daughters

Although the ram was viewed as an important part of naval warfare after Lissa, the gun remained the prime weapon as it had been ever since the Spanish Armada. Soon, however, its position was being disputed by the torpedo.

Originally 'torpedo', a word derived from the Latin for an electric eel, had been used to describe what is now called a mine – a stationary explosive charge resting on or attached to the sea bottom. These were first used in any numbers by the Russians to protect their ports against the British and French fleets during the Crimean War and were effective enough against small enemy ships if they could be brought into contact with them. They were the torpedoes of which Farragut was speaking when, at the battle of Mobile Bay in 1864, it is said he ordered: 'Damn the torpedoes! Full speed ahead!' Whether Farragut actually said this is not clear but that is, in fact, how he acted and when his flagship *Hartford* led the attack her crew could hear the primers of the torpedoes clicking harmlessly as their ship struck the mines, which long immersion had rendered useless.

In the meantime, it was clear to at least one man that what was needed was a way to bring the torpedoes into contact with the enemy, even if his ships did not venture into the minefield. This was originally the idea of an officer of the Austrian Marine Artillery whose name seems to have disappeared, but whose papers, at his death, came into the possession of a retired Austrian naval officer, Fregatten Kapitän de Luppis who, working on the original plans, produced a small wooden boat a few feet long loaded with explosives and driven by clockwork. This was steered by lines from the shore, paid out as the boat headed towards its target.

In that form de Luppis' device was thoroughly impracticable but he took it to an English engineer then working in Fiume at the Stablimento Technico Fiumano, Robert Whitehead. This was in the middle of the great age of British engineering in foreign parts and Whitehead was to become the most famous of them. The British naval historian, Commander E. Hamilton Currey, wrote in 1910:

'No man has ever cost the world so much by an invention as the late Mr Whitehead of Fiume.'

'Whitehead torpedoes' became household words. Whitehead became rich and his children and grandchildren married well, the daughters and granddaughters being known as 'Mr Whitehead's torpedoes'. One daughter married Captain Charles Drury, RN, who afterwards became Admiral and Second Sea Lord, another daughter married Count Hoyos and their daughter married Herbert von Bismarck, son of the German Chancellor. Whitehead's son, James, married the daughter of the 8th Lord Middleton and was the father of Sir Edgar Whitehead, Prime Minister of Southern Rhodesia, while his son John's daughter, Agathe, married Korvetten Kapitän Ritter von Trapp, the most successful of Austria's submarine commanders in World War I and she was the mother of six of the Trapp children who became famous through the film 'The Sound of Music'.

Whitehead received decorations from Austria, France, Prussia, Denmark, Portugal, Italy, Greece and Turkey, but none from Britain. This hurt him and was probably due to Queen Victoria's displeasure at his accepting the Order of Franz Josef from the Austrian Emperor without asking her permission.

Meanwhile he made a fortune, which he spent lavishly, buying a 3,000 acre estate in Surrey and two private Pullman coaches which, attached to appropriate trains, he used for the transport of his friends and business associates around Europe. In all he sold his weapons to eighteen navies, so that there was no naval belligerent from the last quarter of the nineteenth century who was not equipped with Whitehead torpedoes with which they were able to wage war against each other.

Some scandal was caused by the action of a Berlin firm called Schwarzkopf which, after Whitehead's plans had been stolen, placed a torpedo of its own on the market that was, in fact, an improved Whitehead.

It was inevitable that as the navies of the world built up stocks of torpedoes they would soon come into use. The first was used by a British frigate, HMS *Shah,* in action with a hijacked Peruvian ironclad, the *Huascar.* This ship, one of the strongest in the Pacific, had taken the wrong side in a Peruvian civil conflict and, under the command of de Pierola the rebel leader, had sailed up and down the coast of Peru threatening with bombardment towns loyal to the legitimate government if they did not pay ransom to him. The *Huascar* was clearly not under the control of the Peruvian government and was interfering with British shipping. Accordingly the *Shah,* then the flagship of the British Pacific Squadron, endeavoured to deal with the situation. What happened showed how very difficult it was at that time to cope with ironclad warships. For over two hours the *Shah* had the *Huascar* under fire at a range of about 1,000 yards, and hit her seventy times, but only one hit penetrated her armour and that did no damage. *Shah* then tried to use her torpedo; the torpedo sped away into the darkness and nothing more of it was ever seen again.

De Horsey, the British admiral, had laid up trouble for himself. The Peruvians complained that their ship had been endangered by the *Shah*'s attack, and at home in England de Horsey's attempt to use a torpedo was considered as rather unsporting. The situation was made worse by a general belief that the torpedo was a weapon so horrible that it would make the waging of war impossible, a belief which Whitehead seems to have shared, up to a certain point.

Two years later war broke out between Peru and Bolivia on one side and Chile on the other for the possession of the rich nitrate fields which then lay in Peruvian and Bolivian territory. *Huascar* was again active, serving as the flagship of the Peruvian Admiral Grau and this time it was she who fired a torpedo, not a Whitehead but a Lay – a contrivance which apparently had to be fired by a man who controlled its dispatch with four ropes tied to his hands and feet.

On this occasion, in action with the Chilean corvette *Abtao*, the torpedo was fired and its track was marked by flags appearing on the surface. Members of the crew of the *Huascar* watched the progress of their weapon and then saw that it had turned round and was headed back towards them. One of the watching officers, Lieutenant Diez Causeco, jumped into the sea and swam towards the torpedo, stopped it and turned it away. After this the Peruvians lost interest in the weapon and Grau, enraged, had all his stock of torpedoes buried in Iquique cemetery.

It was not until 1891 that the torpedo achieved its first clear success, when a civil war broke out in Chile, the features of which were reminiscent in many ways of the events of 1973 when the government of President Allende was overthrown.

A conflict between a left-wing president, Balmaceda, and a right-wing congress led to civil war. The army, made up mostly of conscript peasants, sided with the president and the navy supported congress, so that on the one side there was an army without a fleet and on the other a fleet without an army. The army tried to rectify this situation by pressing on with the construction of a battleship and two cruisers already ordered in France and, at the same time, took delivery of two torpedo gunboats built in England. These two gunboats, *Almirante Condell* and *Almirante Lynch*, were ordered to be sailed to Chilean waters. However, the civilian passage

crews in the two ships thought that to be too dangerous and refused
to take the ships beyond Buenos Aires. The Balmacedists organised
their onward voyage, but when the ships reached Valparaiso their
machinery had been so badly handled that the boiler tubes were
burned out. The only intact tubes available were those in the boilers
of railway engines. These were removed and fitted in the boilers of
the gunboats, which were then more or less ready for service. And
just in time, from the point of view of the Balmacedists, for Chile
was then almost entirely dependent upon the sea for the movement
of trade and troops, since the Andean geography was such that rail-
ways were only short lines running inland from the coast.

It was clear that the Congressional forces would wish to attack
government forces before the new ships being completed in France
arrived and, accordingly, it was decided that the newly arrived gunboats should attack two important ships reported to be lying at
Caldera. One of these was the *Huascar* which had been hijacked a
second time, having been laid up in reserve, for she was by now an
elderly ship a quarter of a century old. However, when the main
body of the fleet withdrew from Valparaiso, it was decided that the
Huascar should go too and she was recommissioned. Thus on the
night fixed for the attack by the Balmacedist gunboats she was not
in Caldera Bay, where the *Blanco Encalada* awaited her fate alone.
This ship was a battleship of 3,500 tons, armed with six 8-inch guns.
Her departure with the rest of the fleet from Valparaiso had been
so hasty that she had left her torpedo nets behind and this probably
settled her fate. These nets, hung on long booms some distance
from a ship's side, provided some protection in the early days of the
torpedo, but it was not long before torpedoes were fitted with net
cutters and the brief day of the net was over.

On this occasion, however, there was no net. The *Condell*, commanded by Capitano de Navio Moraga, led the way in beneath a
moon intermittently obscured by drifting cloud. *Condell* fired her
bow tube and then turned to bring her beam tubes to bear. All three
torpedoes missed but the *Blanco Encalada,* now thoroughly alerted,
opened fire with everything available, from quick-firers to revolvers.
These weapons hit nothing and diverted everyone's attention from
the *Lynch* which now entered the harbour. Her captain, Commander Fuentes, also fired his bow tube first, at a range of 150
yards, missed and then, copying Moraga's tactics, came closer still
and fired two beam torpedoes. One hit and the *Blanco Encalada*
sank within six minutes, with a loss of nearly 200 men.

After this action it became even more important for the Con-
gressionalists to finish the war before the new warships arrived from
France, and this they did after some stiff fighting, taking an

amphibious force south and capturing Valparaiso and Santiago.

Two years after the Chilean civil war another South American civil war broke out, this time in Brazil, and this involved the use of the torpedo to decide the result. The Brazilian navy, under the command of Rear-Admiral Mello, rebelled against the legitimate government under President Peixoto and occupied some fortified islands in the harbour of Rio de Janeiro.

The fleet threatened a full strength bombardment of Rio but foreign diplomats intervened and a kind of unofficial armistice ensued, which lasted from August 1893 to April 1894, broken only by occasional exchanges of fire between the forts and the ships. In this semi-static confrontation between the two sides the key piece on the board was Mello's battleship *Aquidaban*,[4] which dominated the harbour. Meanwhile Peixoto managed to acquire the rudiments of a navy, consisting of a converted merchantman, a new British-built torpedo gunboat, *Gustavo Sampaio*, and three small German-built torpedo boats.

Of these ships it was the converted merchant ship *El Cid* that attracted the most attention at the time. It was an era of secret weapons, advertised by their inventors as being able to win wars on their own, if not make war so terrible that universal peace would break out. *El Cid* was armed with another of these secret weapons, the dreaded Zilinski pneumatic dynamite gun, which was able, by compressed air, to fire a projectile containing 500 lb. of dynamite a distance of 3,000 yards.

However, it was the *Sampaio* that was to settle matters. This was probably as well since it appears that *El Cid* only carried one of her miracle weapons and various parts of two more. In the Spanish-American War of 1898 the Zilinski gun was fired several times at the earthworks defending Santiago de Cuba without doing any serious damage.

At Rio, on the night of 14 April 1894, the four loyal craft attacked the *Aquidaban*. They quickly became separated and lost their way but *Sampaio*, perhaps by good luck rather than good management, came up to the *Aquidaban* which was giving away her position by the heavy and ill-directed fire of her light armament. *Sampaio*'s captain steadied her on course and came within 450 feet before giving the order to fire his bow tube, only to discover that it had already been fired in the preceding confusion. He then brought the *Sampaio* round the stern of the *Aquidaban* and tried to fire a beam tube by remote control. The remote control gear, however, did not work and the ship's executive officer, rushing back along the deck, fired the torpedo by hand. It hit the rebel battleship, which ceased fire and made for the shore to run herself aground, but sank in shallow water, and was later raised. She was the first battleship to be sunk twice; on 21 January 1906 an internal explosion took place on board and this time she was too badly damaged to be salvaged.

The sinking of the *Aquidaban* really confirmed the place of the torpedo in the arsenal of modern naval weapons. The ship had been the single support of the insurgents and a torpedo had disabled her. From that incident enormous changes in naval warfare were to flow.

During the summer of 1894 unofficial fighting began between China and Japan for possession of Korea, then a more or less independent empire. After six weeks of fairly small scale preliminary hostilities a declaration of war by both sides followed and the two fleets met off the mouth of the River Yalu. Compared with the naval battles between the Russians and the Japanese which were to follow ten years later in these waters this was a comparatively small affair. The Japanese at this time had no battleships, while the Chinese had two slow, well-protected ones. The Japanese fleet was composed of fast, lightly armoured cruisers and gunboats.

When the two fleets met the Japanese were in line ahead and the Chinese in line abeam. Using their higher speed the Japanese steamed around the enemy, keeping up a heavy fire and sinking five of the smaller Chinese vessels, while the Chinese battleships remained almost undamaged. Finally, despairing of being able to sink these ships, Ito, the Japanese admiral, withdrew and the Chinese, their ammunition almost exhausted, returned to their base at Port Arthur.

The Japanese army followed them, landing at Pitzuwu on 24 October and the Chinese fleet left for Wei-hai-wei, the Japanese having made one of their very few mistakes in this war by leaving Port Arthur unwatched. Now, with their army and navy alike, they moved on Wei-hai-wei on 30–31 January 1895, took the town and attacked the Chinese fleet which still lay in the big, wide bay.

Ito decided that it was time to use his torpedo-boats and their torpedo tubes, referred to in those days as 'fish torpedo dischargers'. The Japanese boats were about a hundred feet long and their reciprocating engines were supposed to give them a speed of about twenty knots in calm water, but off Wei-hai-wei that winter there was no calm water. Nevertheless, the little boats pressed home their attacks with great spirit in a kind of dress rehearsal for the attack nine years later on the Russians at Port Arthur. In these attacks the boats, their decks and their weapons were coated with ice, while some men were washed overboard and others literally froze to death at their action stations. Very early on the morning of 5 February, just after moonset, four boats under Togo made a diversion off the western entrance to the harbour and they were followed by the main Japanese force of ten more boats which cut through the obstruction at the eastern entrance. Of these ten boats, two were thrown by the sea onto the rocks, and of the eight remaining only four were able to launch their torpedoes, the others having their tubes blocked with ice or else unable to take aim in the raging seas.

However, one boat did succeed in getting a hit on the Chinese flagship, the battleship *Ting Yuen*, which settled on the bottom of the harbour with her guns still above water and still in action. Three other Chinese warships were sunk, one of them the cruiser *Lei Yuen*, which capsized, leaving part of her crew trapped alive inside. The cries of the entombed men could be heard for several days afterwards, until all was still in the sunken ship.

As the result of the combined attacks from land and sea Wei-hai-wei was now Japanese, at a cost to the navy of 2 torpedo-boats, 29 officers and men killed and 36 wounded.

The Chinese admiral, Ting, a former cavalry officer, surrendered and committed suicide, a brave man and steadfast, despite the heartbreaking behaviour of corrupt authorities who, from the famous Dowager Empress downwards, embezzled money designed for the army and navy and spent it on themselves.

The battleship *Chen Yuen*, sister-ship of the *Ting Yuen*, fell into the hands of the Japanese and served in a minor capacity in the Russo-Japanese war.

By that time the torpedo-boats and their weapons which had fought at Wei-hai-wei had become obsolete and were replaced by much larger and better armed boats. The vessels which had replaced the torpedo-boats were originally called torpedo-boat destroyers, but instead of destroying torpedo-boats they had taken their place. First introduced by the British navy in 1893 they were about 350 tons displacement against 80 tons of the torpedo-boats at Wei-hai-wei, their speed had risen from 20 knots to 30 and they carried 12-pounder guns, as compared with the 1-pounders of the torpedo-boats. Similarly, the torpedoes which they carried had grown and were now of 18-inch calibre instead of 14-inch, with warheads containing 171 lb. of TNT instead of the 88-lb. warhead of the 14-inch. Even more important than the increased power of the latest torpedoes was their ability to steer a straight course and to maintain a set depth. In the Sino-Japanese War torpedoes, when fired, sank rapidly to the bottom or, alternatively, ran on the surface tossed about from wave crest to wave crest. Right through World War II torpedoes, British, US, German and Japanese, often maintained very eccentric standards of behaviour, although in comparison with the early days they were reasonably well conducted.

That this was so was due to the introduction in torpedo design and maintenance of what was known in the British navy as the balance chamber or 'The Secret' which was a development of the gyroscope. Describing the care taken to preserve 'The Secret' Mr Gray writes in *The Devil's Device*: 'The excessive desire for secrecy often degenerated into wild Victorian melodrama,' and he adds 'The ritual of initiation into the mysteries of the balance chamber were described by Lieutenant G. E. Armstrong, RN,[5] in somewhat ironic terms: "The pains taken to preserve this secret were as elaborate as they were futile. The room where the great mystery was unravelled was closed, with locked doors, with sentries on guard outside and every port hole or window carefully screened or closed." '

Writing in *The Cornhill Magazine*, Lieutenant Armstrong added later: 'The "secret" of the balance chamber was kept from the common herd and when at length it was divulged to the budding sub-lieutenant, all the hocus-pocus of closed doors and signed pledge of secrecy was gone through before this ingenious but simply contrived portion of the apparatus was revealed.' This kind of super-secrecy was matched fifty years later by the Japanese who mounted 18-inch guns in the battleships *Yamato* and *Musashi*, but never informed the admiral commanding the ships of the true size of these guns.

In the quarter of the century before the end of World War I the torpedo had had an all-important effect on naval tactics. As its range and accuracy had increased it had forced opposing fleets to stand further and further off lest they found themselves in torpedo range. The gun was still the weapon with the greatest capacity for sinking enemy ships, but it had to do this from outside torpedo range, so that while at Tsushima in 1905 the Japanese and Russians fought at 3,000 yards, the British and Germans fought at ranges between 14,000 and 18,000 yards at Jutland, eleven years later.

It was with ten of the new destroyers that the Japanese began their attack on the Russian Pacific fleet at Port Arthur which opened the Russo-Japanese War on 8 February 1904. There was no declaration of war.

The Japanese attacked in three divisions of four, three and three boats, carrying a total of twenty torpedoes. Of these eighteen were fired. Three torpedoes, all from the first division, hit the battleships *Retvisan* and *Tsarevitch* and the light cruiser *Pallada*. The remaining six destroyers had no success; partly because, at the moment of attack, they had been thrown into disarray by two Russian destroyers which had been patrolling off the port.

None of the three torpedoed Russians were sunk; all were repaired in the next few weeks, thanks to the ingenuity of the Russian naval architects and shipwrights who extemporised cofferdams [6] which made it possible to repair damage done to the ships below the water-line without the use of dry docks, of which there were none at Port Arthur.

The possibility of using a submarine carrying torpedoes to attack an enemy vessel unseen was quickly appreciated, and experiments by the French in the 1890s and the British in the first years of the present century showed it to be feasible, but it was not until 1912 that such an attack was made during hostilities.

In October 1912, a few days before the outbreak of the First Balkan War, the Greek navy took delivery of its first submarine, the 295-ton *Delphin*, built in France by Schneider-Laubeuf. The chief task of the Greek navy at that time was to patrol the exit from the Dardanelles, to prevent the Turkish fleet from recapturing the islands in the Aegean seized by Greece in the first days of the war. One of these islands was Tenedos, on which a base for *Delphin* was established. On 9 December 1912, at about 8.30 am, a Turkish force, made up of the light cruiser *Mejidieh*, the torpedo gunboat *Berk i Satvet* and four destroyers, was sighted emerging from the Dardanelles. *Delphin*, whose captain was Lieutenant-Commander E. Paparrigopoulos, succeeded in taking up a position about 500 yards from the enemy ships and fired her single bow torpedo, the first ever fired by a submarine in war but, although well aimed, it sank before it could reach its target.

4

'Remember the Maine!'
Havana, 1898

On 15 February 1898 the United States battleship *Maine* blew up and sank in the harbour of Havana, with the loss of 266 lives. This disaster was one of the principal events leading to the Spanish-American War.

Spain had been trying to cope with a Cuban insurrection ever since 1868. The United States, close at hand, was sympathetic to the Cuban rebels and extremely interested in the strategic and economic possibilities of the island. In 1878 the first Cuban insurrection was pacified by the agreement of the Madrid government to introduce vital reforms. These were never introduced and eventually in 1895 the insurrection started all over again, temporarily in the presence of Winston Churchill, then a subaltern of the British army who had obtained a long furlough in order to study counter-insurrectionary warfare at close quarters.

During October and November 1897 the Cuban crisis calmed down, due to conciliatory moves by Spain and, as a sign of good will, the Spanish government announced that it was sending the cruiser *Viscaya* to New York on a courtesy call.

In Havana, however, there was rioting between Spaniards and Cubans and in Washington there was the first sign of the seriousness with which the American government viewed the situation. The Secretary of the Navy, John D. Long, ordered the Admiral commanding the European Squadron not to release time-expired men, and it was decided to send the *Maine* to Havana for the protection of American interests in case Americans found themselves the target of either Cuban or Spanish mobs.

The *Maine* arrived off Havana on 25 January 1898, very short notice having been given to the Spanish authorities, but she was allotted a berth in that part of the harbour reserved for visiting warships. She lay there for three weeks, seeming to dominate the harbour by her size in comparison with that of other ships present, although she was less than 7,000 tons. For those days her appearance was impressive, but now she looks to us like a picture of a ship drawn by a seven-year-old, with everything out of proportion compared with ships of the present day. She had two thick, heavy masts, with a large fighting top near the peak of each, and two tall thin funnels. Her hull was white and her upper works and funnels light brown. Her twin turrets, each with two black-painted 10-inch guns, were mounted on either beam so that they seemed to disappear into her upper works when seen from any distance.

The nearest American base to Havana was Key West in Florida and on the night of 15 February the torpedo boat *Cushing*, which had been acting as repeating ship between Key West and the *Maine*, lay there. Late at night a man appeared on board the torpedo boat and asked to see her captain, Lieutenant Gleaves. Gleaves went up on deck and recognised the man as an American agent, who stated that a fellow agent had cabled from Havana reporting that the *Maine* had been blown up. Key West was full of rumours and Gleaves decided to go with the Key West agent to the telegraph office. There they had a wait of hours until the telegraph began to click out a message. It began: '*Maine* blown up in Havana harbour at 9.40 tonight and destroyed. Many wounded and doubtless more killed or drowned . . .' and was signed 'Sigsbee', the *Maine*'s captain.

Out of a complement of 354 officers and men 2 officers and 264 men lost their lives. The very small losses of the officers was due to the fact that the explosion had taken place in the forward part of the ship where the ratings had been berthed.

At daybreak on the morning of 16 February it was seen that the after part of the ship was above water, with the main mast standing upright. The midships was nothing but twisted wreckage standing just above the water, with one funnel lying horizontally on the deck, while the forward portion of the hull had entirely disappeared.

Both the American and the Spanish authorities ordered inquiries into the disaster, but without waiting for the results Americans and Spaniards alike took firm standpoints. To the Spaniards it was unthinkable that any of their nationals would have committed such an act. In the first place the government in Madrid was desperately anxious to prevent the United States becoming involved in the fighting in Cuba, because it was clear they would be on the side of the Cubans and the Spaniards were far from equal to the Americans in resources of any kind.

There was, however, the possibility that Cuban extremists, anxious to involve the Spaniards and the Americans in a conflict from which they would benefit, might have attached a mine to the hull of the *Maine*. In this case the Spaniards would be almost as guilty by reason of their carelessness.

Without waiting for the result of either inquiry public opinion in the United States jumped immediately to the conclusion that the explosion was the fault of the Spaniards and the slogan 'Remember the *Maine*, To Hell with Spain' was the cry of the war-party.

The Spaniards had their report ready first. This declared that their authorities were in no way responsible. Meanwhile, American divers examining the wreck reported that working conditions were extremely difficult; the water was very nearly opaque, the mud on the bottom of the harbour was several feet thick and seemed full of bits of metal which threatened the divers' air hoses and life-lines. This was before the days of skin-diving and the great advances in the art of diving and salvage work it has brought with it. In consequence divers were without the experience required for thorough investigation.

The most spectacular part of the destruction caused by the explosion was in the neighbourhood of the forward 10-inch turret, which had completely disappeared, while the keel in this area had been blown up into a gigantic inverted V, with the angle of the V standing about four feet out of the water, although the harbour bottom, upon which the remains of the ship rested, was 36 feet below the surface. In this area, too, 8-inch armour-plating on the port side of the ship had been blown out. Despite all this upheaval the bow of the ship was still connected to the rest of the wreck.

The night of 15 February had been warm and quiet. Many men had brought their hammocks up on deck to sleep there as an escape from the heat of the mess deck below. At 9.40 pm the ship's executive officer, Commander Wainwright, was talking in the captain's office with a cadet. Suddenly the lights went out, leaving the ship in complete darkness; the door of the office banged shut. There was a heavy shock, a trembling and lurching action of the ship, a roar of immense volume and the sound of objects falling on deck. Wainwright said afterwards: 'I was under the impression from the character of the noises that we were being fired upon.'

Shaken by the violent action and blinded by the darkness it took Wainwright an appreciable time to find and open the door, move out and feel his way to the poop where he found the captain and those officers who had been on deck. Under the captain's authority, writes Wainwright's biographer,[7] 'he did what was needful and possible. He called for perfect silence and heard groans and cries for help from the water on the starboard side.'

Boats had to be lowered to rescue the men in the water but officers, attempting to supervise this, had to man the falls themselves for very few men appeared. Lowering away, the boats became stationary when they had been lowered only a few feet, much less than usual. Unable to see why because of the darkness, they asked the boats' crews when the boats were water-borne, and discovered that the ship had sunk so much that the boats were already afloat. These then began to pick up survivors; in this they were joined by boats from the American liner *City of Washington* and from Spanish warships in the harbour. Wainwright's next reaction was to order the flooding of the magazines, but the sea was already doing that.

In the blackness Wainwright saw that the ship was on fire forward. For an instant, he said later, he took this for 'the wreck of the fire-ship sent down on us'. To find out what was happening he went forward with Cadet Boyd. They found the quarterdeck covered with a mass of wreckage and impassable. The awning was also cluttered with debris, but they managed to crawl forward over it to where the break of the midship superstructure had been – there was nothing there. The whole thing was gone. Nothing remained to starboard from which cries for help could be heard.

Wainwright and Boyd extricated two men and got them into the gig and then returned to the poop to report the 'awful and total' character of the disaster and that there was no hope of putting the fire out.

Wainwright then went in one of the boats to take part in the rescue of the men in the water, working by the light of the fire and of a searchlight trained on the wreck from the shore. When it was clear that all the survivors had been rescued, he returned to the poop, to find that the ship had sunk further still, so that the gig was level with the deck. It was time to abandon ship; there was danger of other explosions and the boats should be clear.

After obtaining the captain's permission, Wainwright gave the order: 'Gentlemen, get into the boats.' With the gig backed against the deck, he urged the reluctant captain to embark saying, 'The only way to get rid of the men is to shove off ourselves.' The captain then followed him into the boat and they shoved off, Wainwright taking with him the ship's cat.

Another cadet, W. T. Clutherius, made a succinct and accurate report. He said: 'My first knowledge of anything occurring was a slight shock as if a 6-pounder gun had been fired somewhere about the deck. After that there was a very great vibration in my room which was then followed by a very heavy shock and still continued vibration and rushing of water through the junior officers' mess room.'

Two men in the superstructure near the centre of the ship saw a red flame and then, they said, 'The deck trembled and in a blaze of fire we suddenly found ourselves below, with the water pouring in on us and with a heavy weight on top of us. A second explosion seemed to lift the weight and we escaped with burns.'

From the berth deck only two men were saved. One of them said: 'I heard a terrible crash – an explosion, I suppose it was. Something fell and then after that I got thrown somewhere in a hot place. I got burned on my legs and arms and I got my mouth full of ashes and one thing and another.'

Below the armour deck only one man survived, a stoker. He was lying down in the steering engine room, two decks below the berth deck and right astern. He could see through the open door into the engine room when, all at once, 'a blue flash shot by the engine room lamp'. There was a continuous trembling of the ship and a terrible report. It seemed as if 'the whole earth had opened up'. A man who was with him died in the explosion. Afterwards it was established that men were trapped in closed compartments and could not be rescued as the ship gradually filled with water, while debris burned all that night with frequent small explosions.

The report of the American inquiry into the disaster was ready for President McKinley on 25 March and flatly contradicted the previously published Spanish report. Each upheld the standpoint of its respective government.

War between the United States and Spain began on 25 April and lasted until 13 August. At sea two weak Spanish cruiser squadrons were destroyed and Spain lost what had remained of her possessions in the Pacific and the West Indies. After this there was, naturally, no further Spanish interest in the question of responsibility for the loss of the *Maine,* but in the United States there has been continued spasmodic interest until the present time.

The first impartial investigation of the disaster came as early as May and June 1898 with the publication in *Engineering* of two articles by Lieutenant-Colonel J. T. Bucknill, RE, who disagreed with the findings of the American inquiry and said that the two explosions had been caused by the spontaneous combustion of bituminous coal in one of the ship's bunkers alongside a magazine. It was the practice, at this time, to place coal bunkers in such positions as an additional protection against shells and torpedoes, in spite of the fact that there had been a number of cases in which it was known that spontaneous combusion of coal had occurred.

In 1911 the after-part of the *Maine* was raised, towed out to sea and scuttled in sixty fathoms. However, interest in the *Maine* continued and there was further investigation of the cause of the disaster. As recently as 1976 Admiral Hyman G. Rickover, USN, celebrated for his work on the development of the nuclear submarine, wrote a book called *How the battleship 'Maine' was Destroyed*. This was a re-examination of all available evidence, ending with the conclusion: 'In all probability, the *Maine* was destroyed by an accident which occurred inside the ship.'

The admiral criticises the American investigation of 1898 for having failed to make use of all the resources at its disposal. Had it not been for this failure, he says, there would have been 'the injection of reason into an atmosphere of emotion. At least the United States would not have found itself adopting an official position which was technically unsound and which increasing numbers of people have questioned over the years. And – although the chance was slim – the war might have been avoided.'

Admiral Rickover insists that the affair of the *Maine* has a lesson of vital importance for us today:

In the modern technological age, the battle cry: 'Remember the *Maine*!' should have a special meaning for us; with almost instantaneous communications that can command weapons of unprecedented power, we can no longer approach technical problems with the casualness and confidence held by Americans in 1898. The *Maine* should impress that technical problems must be examined by competent and qualified people, and that the results of their investigation must be fully and fairly presented to their fellow citizens.

With the vastness of our government and the difficulty of controlling it, we must make sure that those in 'high places' do not, without more careful consideration of the consequences, exert our prestige and might. Such uses of our power may result in serious international actions at great cost in lives and money – injurious to the interests and standing of the United States.[8]

These words of Admiral Rickover seem clearly relevant to the unending discussion of nuclear weapons.

5

Three Flagships

Petropavlovsk, Kniaz Suvarov, Mikasa, 1904-5

On the morning of 8 February 1904, after the Japanese destroyers had attacked the Russian ships lying in the entrance to Port Arthur, the Japanese Commander-in-Chief, Vice-Admiral Heihachiro Togo, appeared off the Russian base with his battleships and shelled it without, however, doing any damage or making any attempt to press home his attack. For this he was criticised afterwards but the respective strengths of the Russian and Japanese fleets demanded caution. At that time Japan had six battleships, while the Russians had seven in the Far East, and two more complete fleets in Europe, one, with eight battleships, in the Baltic and the other, with nine battleships, in the Black Sea.

It was true that the Black Sea Fleet was barred from passing through the Dardanelles by the Treaty of Berlin, but the lives of treaties were as uncertain then as they are today, and in any event a combination of the Baltic and Pacific Fleets gave the Russians a superiority of more than two to one. Accordingly the Russian Admiralty, then in St Petersburg (now Leningrad), began preparations after several weeks to sail the Baltic ships to the Pacific, but it was not until nine months had elapsed that they were ready to leave on a voyage which was to take them to their doom after eight months spent steaming half way round the world.

As for the Japanese, as soon as the war began and despite the adverse balance against them, they started to move their army overseas to invade Russian-held Manchuria via Korea.

It was not surprising that immediately after the Japanese attack which had started the war the Russian high command had replaced the commander of the Pacific Fleet, Vice-Admiral Stark, by another Vice-Admiral, Stephan Ossipovitch Makarov. He was one of the very few senior Russian officers of humble birth, both his grand-fathers having been NCOs and he himself had been sent to a naval school in Siberia at the age of ten.

Commissioned, he distinguished himself in the torpedo-boat war with the Turks in 1877–8 and then became a very experienced Arctic explorer, but his first concern was his career as a naval officer. Hoisting his flag in command of the Pacific Fleet he took it in hand, restoring morale and efficiency by means of intelligent activity. An indication of his success was that it took two and a half hours to get his twenty ships out of the complicated exit from Port Arthur, instead of the twenty-four which had previously been required.

Makarov was a very great admiral, but unfortunately for him and for Russia he was confronted by another just as skilful, and in command of a much more efficient fleet.

Vice-Admiral Heihachiro Togo was born in 1847 and had begun his career in the Japanese civil strife of the 1860s, a soldier-Samurai in an army dressed in silken armour (affording excellent protection against sword cuts) and armed with swords and bows and arrows. Once this fighting was over, the Japanese began building a navy. To provide officers for the new Service promising young men were dispatched to Europe to obtain a nautical education. Togo was sent to the *Worcester*, a training ship for officers of the British merchant service, moored in the Thames and, after seven years abroad, including a spell as an ordinary seaman in sail and a course in mathematics at Cambridge, he returned to Japan a fully qualified naval officer.

In Japan's first more or less modern war against China in 1894–5, Togo was a cruiser captain and nine years later commanded his country's battle fleet, surviving the victorious Russo-Japanese War until 1934 when he died at the age of eighty-six.

Makarov's career on the other hand as Commander-in-Chief was to last only one month. Five weeks after taking command his flagship *Petropavlovsk* struck a Japanese mine off Port Arthur, blew up and sank, taking with her Makarov and 635 members of her crew. Commander Semenov, then serving in the cruiser *Diana*, has left a classic account of this disaster in his book *Rasplata* (The Reckoning) published in 1909.

An explosion, with a dull rolling sound, shook the whole ship as if a 12-inch gun had gone off quite close. I looked round vaguely. A second explosion even more violent! What was happening? Suddenly cries of horror arose: 'The *Petro-pavlovsk*! The *Petropavlovsk*!'

Dreading the worst I rushed to the side. I saw a huge cloud of brown smoke. 'That is pyroxiline, therefore a torpedo', passed through my mind. In this cloud I saw the ship's fore-mast. It was slanting, helpless, not as if it were falling but as if it were suspended in the air. To the left of this cloud I saw the battleship's stern. It looked as always, as if the awful happenings in the forepart were none of its concern. A third explosion! White steam now began to mix with the brown cloud. The boilers had burst. Suddenly the stern of the battleship rose straight in the air. This happened so rapidly that it did not look as if the bow had gone down but as if the ship had broken in half amidships. For a moment I saw the screws whirling round in the air. Was there a further explosion? I don't know.

It appeared to me as if the after-part of the *Petropavlovsk* (all that was visible of her) suddenly opened out and belched forth fire and flames, like a volcano. It seemed even as if flames came out of the sea, long after it had closed over the wreck.

When I saw the explosion, I mechanically looked at my watch, and then wrote in my notebook:

'9.43 – explosion on board *Petropavlovsk*.

'9.44 – all over.'

Seven officers and seventy-three men were saved out of a complement of 715. There was no trace of Makarov.

Panic firing broke out aboard most of the Russian ships. When order was restored they returned to Port Arthur, the battleship *Pobieda* having been damaged by a mine.

A month later the Japanese suffered a series of heavy blows. On the evening of 14 May, in foggy weather, two of their cruisers collided and one sank. Next morning three battleships, *Hatsuse*, *Yashima* and *Shikishima*, were patrolling off Port Arthur, when the *Hatsuse* and *Yashima* both struck mines and the *Hatsuse* then struck a second one. There was a great column of yellow flame and the main mast and one funnel, still pouring smoke, could be seen in the air as the ship sank. The *Yashima* sank later while under tow, leaving four Japanese battleships to face six Russians. These were the most favourable odds which the Russians were to have at any time during the war, but they failed to take advantage of them, partly because the Japanese succeeded in keeping secret the loss of the *Yashima*.

Meanwhile, ever since the outbreak of the war the Japanese army had been attacking Port Arthur, hoping to take it before the Baltic Fleet (now renamed the Second Pacific Fleet) could bring help from Europe.

By August, seven months after the war had begun, it seemed unlikely that Port Arthur would hold out long enough for help to reach it and so it was decided that the Port Arthur fleet should try to break out and move its base 1,170 miles northward to Vladivostok. On 10 August the Russians sailed from Port Arthur and Togo and the Japanese fleet pounded along in pursuit. As Togo tried to get ahead of the enemy and cut their escape route two 12-inch Japanese shells struck the *Tsarevitch*, the Russian flagship, killing Vitgeft who had succeeded Makarov as Commander-in-Chief, and killing or wounding everyone else on the bridge. The *Tsarevitch*, which sheered out of line, started to turn in a great circle, followed by her next astern, the *Retvisan*. These two ships continued their turn until they had headed back to their own line through which they passed, narrowly escaping collisions with the *Peresviet* and *Sevastopol*, fourth and fifth ships in the original Russian line.

The senior unwounded officer in the *Tsarevitch* was at his action station below deck and, as his ship continued to circle out of control, he had to be found, have the situation explained to him, take over the ship and report to the Russian second-in-command, Rear-Admiral Prince Ukhtomsky, in the *Pobieda*. However, the *Pobieda*'s signal halliards had been shot away so that Ukhtomsky was unable to give any orders for the German-pattern Russian wireless telegraphy was very nearly useless.

Once Togo saw the disarray of the enemy and that they were now headed back to Port Arthur he did not interfere with their escape. Not all the Russian ships were able to reach Port Arthur. The *Tsarevitch* took refuge in German-held Tsingtau where she was interned, the cruiser *Askold* and a destroyer were interned by the Chinese and the *Diana* was interned by the French at Saigon.

The Japanese ships, and especially Togo's flagship the *Mikasa*, had been damaged but were afloat and able to resume the fight when the Russian Second Pacific Fleet finally reached Japanese waters nine months later.

The task of actually sailing the Second Pacific Fleet around the world had been enormous; nothing of the sort had ever been attempted in the age of steamships, and no attempt has ever been made to repeat it. It was unique; in the first place it was carried out in the middle of a war by one of the belligerents, obliged to respect, for the most part, the neutrals whose ports it was necessary to visit to obtain supplies. At this time there were few hard and fast rules in international law laying down the length of time belligerents might spend in a neutral port, and neutrals tacitly made their own rules. Thus, the Russian fleet did not even request permission to visit a British port but made two long stays at French colonial ports in Madagascar and Indo-China, while the government in St Petersburg debated what should be done after the fall of Port Arthur.

Then, after two and a half months at Nossi Bé in Madagascar –
a terrible port on the edge of the jungle, steeped in hypertropical
heat and with no facilities for the crews – there came the long haul
across the Indian Ocean, during which the ships could only re-
plenish with coal at sea. This was done from chartered French and
German colliers which transhipped their cargoes of coal in bags.
These were placed in ships' lifeboats to be rowed across to the ships
for which the coal was destined. The coal sacks would then be
hoisted aboard. This would have been difficult enough with trained
crews, but these were lacking and the heat and the glare on the
open sea were horrible.

The four brand new battleships of 13,000 tons which comprised
the main body of the Russian fleet were so new that they had never
succeeded in outgrowing their 'teething troubles' and their crews
had never shaken down. Nevertheless, when it came to battle eight
months after the start of this grim journey, the showing made by
the best of Russian battleships was most distinguished and deserving
of a permanent place in naval history. These ships were the
Imperator Alexander III, *Borodino* and *Kniaz Suvarov*, the latter
flying the flag of the Vice-Admiral Zinovi Petrovitch Rozhestvensky
who through thick and thin led his fleet to its destruction. His bearing fluctuated between great rages, in the course of which he regularly reduced his flag captain to tears, and dogged despair as he
sailed on through breakdowns, minor mutinies, diplomatic wrangles,
as well as constant harassment by the Admiralty in St Petersburg
when he wished to use neutral facilities, and a clash with Britain
which very nearly caused a war between the two countries. This
was occasioned early in the voyage, when the Russian fleet entered
the North Sea and opened fire on some British trawlers, although
there were no Japanese warships within ten thousand miles or so.

It was on the night of 26 May 1905, eleven months after the
dispatch of the Russian fleet from Europe had been decided, that it
entered the Straits of Tsushima between Japan and Korea. Six
hundred miles ahead lay the sole Russian naval base in eastern
waters since the surrender of Port Arthur on New Year's Day 1905.

To steam that distance unobserved was vital to the success of
Rozhestvensky's voyage, but it was typical of the mismanagement
that had characterised the Russian conduct of the war that while
the warships of the fleet were all darkened the two hospital ships
accompanying them had all lights burning in accordance with the
regulations for hospital ships laid down by the Hague Convention.
It was these ships which were sighted by a patrolling Japanese
auxiliary cruiser, the *Shinano Maru*.

Togo was waiting for them at his base at the end of the Korean peninsula among remote islands reminiscent of Scapa Flow. On learning that the Russians had been sighted he sailed at once, leading his ships almost as far east as the coast of Japan, then finding that he was too far north turned back again more or less on the course along which he had come. Almost immediately he found the enemy and began to steam across the somewhat irregular Russian line, thus achieving the manoeuvre known as 'crossing the enemy's T' which enabled the Japanese battleships to take as targets each one of the enemy ships as they crossed the column, while only the *Kniaz Suvarov* at the head of the Russian line would be able to reply as she was masking all her consorts. On the other hand, the Japanese were at considerable risk as their battle line had, ship by ship, to pass through a fixed point at which each Russian ship could aim.

By far the best account of the battle that followed, from the Russian viewpoint, was that written by Semenov from the *Diana* in two books, *Rasplata* (The Reckoning) and *The Battle of Tsushima*. After his ship had been interned by the French at Saigon he had escaped from their not very strict surveillance and was now attached to Rozhestvensky's staff in an undefined capacity. Now, with the enemy in sight, Semenov took up his position on the quarterdeck of the *Suvarov*, pencil and notebook in hand.

The first direct hit on the *Suvarov* started a serious fire, and Semenov noted that the damage-control party nearby stood watching the flames in a state of shock until he quietly told them what to do, when they fell to with a will. Then it was Semenov's turn. He felt a heavy blow in the back and the next thing he knew he was on the deck looking for the glass of his watch, which was going normally except that it had lost its minute hand. As he collected his wits he saw that there were two or three bodies on the deck, while the fire hoses which lay over the deck were spraying water at random, having been badly cut up by shell splinters.

Whilst watching the Japanese shells exploding Semenov realised that their fuses were more sensitive than any shells he had seen before. They burst on contact with anything they hit – wire rigging, davits, ships' boats – and shells bursting on plating generated such heat that the paint on the steel plates caught fire.

At the same time the Russians, able to watch the effects of their shooting, did not have any corresponding satisfaction. A large proportion of their armour-piercing shells failed to explode, while others which did pierce the Japanese armour burst inside the enemy ships and were almost invisible at any distance.

The *Suvarov* replied to the Japanese fire at 1.49 against the two battleships leading the enemy line – *Mikasa* flying Togo's flag, and *Shikishima* – at a range of 6,400 yards. These were followed by the

other two Japanese battleships, *Fuji* and *Asahi*. It had taken a very
short while for the Japanese to find the range; a few rounds over
the *Suvarov*, a few short, the splinters from which rattled inboard.
Then, abeam of the forward funnel, a great column of water
mingled with flame and smoke rose up and stretcher bearers ran
forward along the deck. Semenov found that the first shell had
burst in a temporary dressing station killing the chaplain and all the
sick berth attendants, but leaving the doctor unhurt. The ikon there
remained intact – its glass cover unbroken with two candles still
burning in front of it. The next shell hit abeam of the midships
6-inch turret on the port side and long flames emerged from the
officers' quarters. A third shell burst outside the conning tower in
which were Rozhestvensky and his flag captain, the former crouch-
ing down to see out of the sighting slit between the wall and the
roof. Behind them, one on each side of the wheel, lay two officers
face down on the deck. The range had now fallen to 4,000 yards
and through his glasses Semenov could see the enemy on the decks
of their ships with their hammocks hung up as anti-splinter mats.
Looking about his own deck Semenov saw groups of dead and badly
wounded. Damage-control positions had been destroyed and a
number of small fires had been started. Behind the *Suvarov* came
the *Borodino* and *Alexander III*, both partly hidden by clouds of
smoke from the fires which were blazing aboard.

It was five minutes past two and the battle had, in fact, been
decided. There could be no Russian recovery from the state in which
they found themselves, although the agony was to be prolonged until
the morning of the next day. A messenger reported to Semenov
that the after 12-inch turret had been hit. He went aft and found
that part of the turret roof covering the right-hand 12-inch gun
had been blown off, despite which both guns were still in action.

Fire-fighting was slowed down by a lack of unwounded men and
hoses which lay about the ship cut open by shell splinters and leaking water where it was not needed, while the ship's superstructure
was on fire. During the long voyage from the Baltic many men had
spent their time making emergency fire buckets out of old oil tins
and the use of these now postponed the end of the ship.

A midshipman from one of the 6-inch turrets emerged and offered
Semenov a drink of cold tea which, he wrote afterwards, helped him
to pull himself together. He went back to his original position on the
upper deck and looked at the enemy who seemed untouched and as
if they were simply exercising – the continuous thunder of the
Russian guns apparently having had no effect on them.

At about 2.20 pm Semenov received the report that all the
halliards had been cut and it was therefore impossible to make any

signals. No one seems to have considered using radio for this purpose, perhaps not surprising in view of the performance of the German-built Russian gear up to that time. It will be remembered that at the battle on 10 August the *Pobieda*, which led the line after the *Tsarevitch* had been knocked out, was also unable to communicate with the ships under her command.

There seems to have been no thought of using semaphore in either case. Semenov went back to the conning tower which now contained five or six dead men, while Vladimirsky, the Fleet Gunnery Officer, took the wheel, an imposing figure, his face covered with blood but with upturned, bristling moustachios still defiant.

At this moment, for reasons unknown, the rest of the roof of the after-turret fell off sideways, with a great clang, followed by an even heavier crash, and the whole of the upper deck was covered with evil-smelling smoke and bits of burning wood. In a few moments it was possible to see that the forward funnel had collapsed. A collection of panic-stricken men surged around and, with difficulty, were restored to order.

It was now 2.30 and the battle had been decided although fighting was to last into the afternoon of the next day.

The *Suvarov* was hit again, her rudder jammed, and she started off in a circle, while ships ignorant of her condition, trying to follow her, could not do so because of the smoke which hung low and heavy over the sea in great thick clouds.

The first Russian battleship to be sunk was the *Osliabia*. At about 3.30 she was hit forward almost simultaneously by three 12-inch shells from the *Asahi* and sank so quickly that there was no time to open the hatches in the armoured deck that cut off the engine room complement from a chance to save their lives.

Orders had been given before the battle had reached a crisis that two destroyers, the *Biedovi* and *Buistri*, should be ready, if the *Suvarov* was disabled, to take off the Admiral and his staff and transfer them to another ship. The urgent need for this had now arisen, but it was not possible for the *Suvarov* to communicate with her attendant destroyers.

Finally the attention of the *Buini* was attracted by an officer signalling with his hands and she came alongside the flagship. Neither ship had a boat intact. Eventually one of the destroyers came alongside and, despite the scend of the sea, was able to remain close enough for the admiral, now unconscious, to be bundled down the side of his sinking flagship and taken aboard the destroyer.

Two Japanese sources have described the *Suvarov* at this moment. Vice-Admiral Kamimura, Togo's cruiser admiral, wrote: 'Her upperworks were riddled with holes and were hidden by smoke, her masts had been shot away and her funnels as well. She could no longer steer and the fire on board spread with terrifying speed, but although she had fallen out of the line she continued to fight on.'

While the Japanese Official History recorded: 'In this tragic situation, she was still the flagship. She fought on and continued firing with the few guns that she had left.'

The *Borodino* and *Imperator Alexander III* were nearly as badly damaged but were still able to attempt to fight their way through to Vladivostok. The time was now 4.15 and it seemed there might be a hope, although a meagre one, that night might save these blazing ships. Now the Japanese battleships drew off and left the field to their torpedo craft. Togo, like Jellicoe and Beatty at Jutland, did not wish to risk his heavy ships in a night action. The Japanese destroyers, however, settled matters. *Suvarov* was found by destroyers and was quickly dispatched with nine torpedoes. She turned over and floated upside down for a few moments while some men gathered on her upturned bottom. Then she disappeared and the Japanese managed to save some twenty men, survivors from a complement of 750. The British naval historian H. W. Wilson commented: 'Thus ended the most determined action ever fought by a battleship.'

Next morning the *Buini* with the admiral still on board ran out of coal. Rozhestvensky was moved to another destroyer which shortly afterwards surrendered to an overwhelmingly strong Japanese force, as had the remnants of the Russian battle fleet, *Orel, Apraxin, Ushkov* and *Nikolai I.*

Less than four months later *Mikasa*, Togo's flagship throughout the war, was the centre-piece of a great naval victory celebration at Sasebo. On 10 September Togo left the *Mikasa* to go to Tokyo, and on the same night fire broke out on board. Raging violently despite all efforts to check it, it reached the after 12-inch magazine and a huge explosion took place. The ship sank at once in six fathoms and 599 men lost their lives.

Divers, examining the wreck, reported a hole eighty-one feet long in the stern, as well as nine lesser holes. Even less was known then than today about the behaviour of high explosives and there were plenty of rumours and gossip, which were not entirely dissipated by the official government statement giving 'spontaneous combustion' as the cause of the explosion. It was a known fact that bitter ultra-nationalist feeling prevailed against the terms of the Treaty of Portsmouth, New Hampshire, which had ended the war, and which were much less favourable to Japan than the Japanese people considered their due.

Salvage teams worked for nearly a year to raise the *Mikasa*, making her hull watertight and pumping it out, while her two twin 12-inch turrets were removed to lighten her. Fires were then lit in her boilers, steam raised and the ship steamed almost literally from the bottom of the harbour into dry dock where she was refitted. In 1922 she was spared by the Washington disarmament conference and demilitarised, remaining as a national war memorial.

In 1945, at the end of World War II, a wave of pacifism swept Japan. National war memorials went out of fashion and the hull of the *Mikasa* served as a kind of amusement centre, with an aquarium and dance hall. The ship's third escape came in the early fifties when she was entirely rebuilt and reappeared in her original condition, once more a war memorial and a museum, and was opened to the public on 27 May 1955, fifty years to the day after Tsushima had been fought.

6

Internal Explosion

Iéna and *Liberté*, 1907-11

On 12 March 1907 the French battleship *Iéna* was in dry dock at Toulon. Inboard and outboard all the upsets, disturbances and inconveniences which inevitably accompany a stay in dry dock for every ship were in full swing. It was noisy, all the hammering and drilling in the world seemed to be going on, and the men responsible were spread all over the ship, on decks, in cabins, in alleyways and here and there in parts of the engine room and boiler rooms. Men who were not making a noise were engaged in slapping on paint. There was no water for washing or cooking and no plumbing.

Sub-Lieutenant (*aspirant de 1ère classe*) Nepveu looked at all this activity and all these inconveniences with distaste, but reflected that it would soon be over and the ship would be to rights again and sailing with the other two battleships of the second division of the Mediterranean fleet. He went into his cabin and resumed work on calculations that he was making in connection with a range-finding exercise.

Suddenly he felt a blast of air, which sprang up unexpectedly and brought with it stones that rattled against the walls of the dry dock. Through both the scuttles of his cabin he could see and feel violent gusts of wind, blowing in more pebbles and small stones. Then the great ship moved with a jolt which made him think that she had lurched forward, off the baulks of timber on which she was resting, and hit the end of the dock.

'I sat bolt upright in my chair; a whirlwind seemed to enter the cabin through the scuttles, knocking me down head over heels, and tearing down a locker and the washbasin from the bulkhead.' Three minutes later he came to, with a cut on his head which was bleeding a lot and another on the bridge of his nose.

The electric light went out suddenly. He thought that probably the explosion had occurred near the dynamos of one of the magazines, either the 138.6 mm or the 305 mm next to it; already there was a bitter smell of cordite which caught his throat and seemed to confirm that impression.

Stepping over the wreckage I seized the knob of the metal door. The door was jammed in its frame and I could not open it. I went to the two scuttles which were darkened by smoke, white, green and black, already leaking into the cabin. Perhaps I would be able to get out through their 350 mm diameter. I tried hard and managed to get about half my body through when I had to go back into the cabin again. Above and below me the side of the ship was on fire and the ladder leading to the ward room, down which I had come a few minutes before, was on fire. In the smoke I saw a messmate, who saw me and raised his arms to heaven in a moment of desperate helplessness. That was why my name was added to the list of dead. 'They'd seen me burn.'

It was 1.02 pm and I believed I was going to be burned alive, suffocated or be blown up in one of the explosions which I could hear every minute killing more people and adding to the confusion.

I turned back to the jammed door. I crashed into it, using my body as a battering ram, once, twice, nothing happened and my shoulder was already most painful. I attacked it again at different points, each blow more successful, three times, four times; at the top it was beginning to give. At last five or six more attacks and the door did give and I fell into the dark of the alleyway, as the deck fell from under me and one of my legs dropped into a bottomless hole. I realised what had happened. There was a coal chute in front of my door and the first explosion had blown the lid off it like a cork out of a champagne bottle.

Nepveu stood up painfully, his leg was hurting a great deal. Two steps away was the fire door of the wardroom ante-room. His first thought was to open it. The fire in there was burning fiercely and perhaps, behind that door, there were wounded and dying men. He paused, remembering that the fire doors had to be kept shut. That might save the ship. He dared not open them because he was not sure of being able to close them again. He felt his way forward in the dark to the ladder leading to the mess decks, a vast area generally full of people busy in different ways according to the time of day – men under instruction, men going ashore, men in working parties – but now the big compartment was entirely empty. There was a dim half light greenish-white, green and black and red and yellow as well, coming from the door on the starboard side from which a gangway went up to the side of the dock. But that gangway was on fire and already half burned away. He hesitated, interrupted by more explosions aft. The fire was dying down a little, the gangway seemed still solid. Three quick strides and he was on the dockside. The breath of the fire licked his face, but the cloth of his jacket did not catch fire.

There was no one on the dockside. Then, coming out of the smoke, he saw Lieutenant Roux. 'I am going to flood the dock,' said Roux, 'you go and report to the admiral who is at the head of the dock.' Roux hurried off aft. A little later there was another explosion which blew off both his legs, killed him and threw his body to the bottom of the dock.[9] Nepveu went on towards the head of the dock. Looking back he could see that the fire was fiercer than ever, giving off torrents of smoke, and he wondered why the after 12-inch turret had not blown up. It would soon.

At five minutes past one Nepveu met the first party of men he had seen since the explosion had begun. Clinging to the ropes which ran from the bow of the ship to the dock, they were making their way hand under hand to safety and then found themselves contending with hot tar; now, one by one, they fell thirty feet into the bottom of the dock, where more than ten of them already lay.

Right forward an admiral, some officers and about forty men were watching helplessly. Nepveu reported what Roux had been trying to do and said that, in his opinion, the 12-inch after-magazine had not exploded but that the stern of the ship would soon be destroyed by an explosion even greater than the ones which had already taken place. When that happened the open space where they were standing would receive a torrent of shells.

Then, as he heard the order to clear the space, he collapsed. When he came to a few moments later, at 1.10 pm, he was on his way to the hospital. There was a tremendous explosion and, turning his head, he saw very high in the air huge pieces of wreckage. The after-part of the *Iéna* had blown up.

There was still great danger to be feared from the blazing ship, and the dock had not yet been flooded. The wheel which controlled the flooding needed more than twelve turns before water would flow into the dock and extinguish the flames, but explosions and confusion made this impossible. Accordingly the battleship *Patrie*, 500 yards away, trained a 12-inch gun on the dockgate and fired, smashing it and letting the water in. Meanwhile men from the yard, some burned and all smoke-blackened, had rushed out into the streets of Toulon, where they were rounded up as firmly as might be and interrogated as to what had happened, for the authorities and the citizens knew nothing more than that there had been a series of explosions and a great fire, which was still burning fiercely. The interrogation proved a hopeless task as all the men were in such a state of shock that they were unable to speak.

In all, 117 men lost their lives. As is always the case in a disaster there were some extraordinary escapes; perhaps the oddest among the survivors was a group of stokers who owed their safety to the fact that when the fire started they were, somewhat improbably, attending a lecture on Henri IV given by a sub-lieutenant whose audience, about fifty strong, was able to escape over the bows of the ship. The inquiry into the causes of the disaster pointed to the notorious 'Poudre B', which had already nearly destroyed two ships, the battleship *Amiral Duperré* and the sloop *Forbin*, because of its instability, but these vessels had been saved by timely flooding of their magazines.

Four and a half years after the disaster to the *Iéna* a similar catastrophe befell another French battleship at Toulon, the *Liberté*, at 5.30 in the morning of 29 September 1911. Seventy years afterwards a hitherto unpublished account of it was sent to the writer by the kindness of Commandant Jean Claude Bertault and Monsieur Bertrand Dousseau of St Germain-en-Laye.

Written at the time by Lieutenant de Vaisseau Robert de Dreuzy, then Commanding Officer of the destroyer *Voltigeur*, it begins with a narrative of events in the *Liberté*, given to Dreuzy by one of her Petty Officers:

It was about 5.35 am. At 5.30 the men had turned out and, still half asleep, were stowing their hammocks on the mess decks. Petty Officers, in their mess alongside the forward batteries, were drinking their coffee.

Suddenly, below them, they heard the crackling of many small explosions. For a few seconds there was nothing more and then, that same crackling again. At once vivid yellow flames and choking yellow smoke came up from below on to the mess decks. The men, surprised while still stowing their hammocks, tried to seek shelter. Others copied them and it seemed as though the whole ship's company was aft, where the Officer of the Watch was superintending the hoisting in of the boats. Seeing at once the extent of the danger, he ordered the fire alarm sounded, but in the confusion the bugle could not be found.

At this moment the senior officer on board, Lieutenant Garnier, came on deck, ordered the crew to their fire fighting stations, hoisted the signal for 'Immediate Assistance' and told

the ship's Chief Engineer, Mécanicien Principal Lestin, to flood
the forward magazine.[10]

Lestin, with the Chief Gunner, made two attempts to flood
the magazine and twice they were obliged to return to the upper
deck. They could not reach the wheels which controlled the
valves letting water into the magazines, now in the midst of
fire, smoke and explosion.

In addition, ever since the first explosion, electricity had
failed throughout the ship and the 'tween-decks were in complete darkness.

Desperate, Lestin returned to Garnier and reported his lack
of success.

Garnier put his arm around the Engineer's waist and, stress-
ing every word, said:

'Wretch! Get this clear! Go back at any cost!'

Lestin, having understood, saluted formally and disappeared,
followed by the Gunner, and neither were ever seen again.
This third attempt to reach the valves cost them their lives.[11]

Like Roux of the *Iéna* Lestin was posthumously honoured by
having a destroyer named after him.

While this was going on, Dreuzy recalled:

Something happened which I cannot praise too highly.
When the fire broke out many jumped overboard. When they
heard the bugle calling them to 'Fire Stations', they began to
swim back to the ship, while others who had already been
picked up by ships nearby had jumped overboard again and
swum back to the *Liberté*.

Most of these men are dead now . . .

The first rush of flames had set fire to the bridge and the foremast
but it had subsided, having started a number of small fires which
showed red through the smoke, and fires were also reported in the
56-mm and 47-mm magazines. However, at 5.43 am although the
Liberté could be heard sounding a general alert and was still flying
the signal for 'Immediate Assistance', the danger seemed to have
diminished. In order to drive the smoke over the bows of the
Liberté a tug was summoned and began to pull the battleship's
stern into the wind.

Suddenly the flames and the smoke disappeared, so that it
seemed that Lestin had been successful. The waiting boats from
the other ships of the fleet approached the *Liberté*. An officer from
one of them hailed Bignon, of the *Liberté*, and asked whether

he should come alongside and begin to take off her people.
'No,' replied Bignon. 'It's dying down.'

At that moment an enormous explosion occurred. When the
dense smoke had cleared away, it could be seen that the forepart of
the ship had entirely disappeared. The upper deck of the midship
section had been peeled back so that the 7.6-inch turrets mounted
on it, although still attached, were now upside down, while aft, the
12-inch turret was level with the water and the ensign staff at the
stern of the ship was still standing upright. The forward part of
the ship rose up vertically and then fell back on the stern. A mass
of wreckage weighing 40 tons was thrown 150 yards, crashing down
on the battleship *République* abeam of her after-turret, and all over
the harbour wreckage was falling and shells were exploding.

The body of the *Liberté*'s bugler, with the bugle which had been
missing still in his hand, was also found on board the *République*.

Dreuzy gave his story in a letter to a friend:

I was awakened in my flat by the bells of the Arsenal sounding the Alert. I could see from my window that there was a
fire in the roadstead and the brilliant rose colour of the flames
made me think of Powder B. I dashed down to my ship and
found the crew rather shaken. I tried to reassure them as
best I might.

I selected the best swimmers and divers for the whaler and
the dinghy, together with my sick-berth attendant and placed
mattresses at the bottom of the whaler.

As soon as we got outside the port I saw that the situation
was desperate. The fire was raging and from all over I could
hear long drawn out cries of agony.

Never before had I heard men dying like that. It was
frightful and I admit that I was trembling. I was about a
thousand yards from the *Liberté* when the explosion took place
with a tremendous noise. Then everything was completely
dark, darkness broken by cries and shrieks as enormous objects
falling into the water all around us, covered us with splashes and
splinters which we could not see to avoid. When the smoke
cleared away I saw things unforgettable, and where the *Liberté*
had been there was a shapeless mass.

Cries of pain came from all over, the roadstead was strewn with
wreckage of all kinds to which wounded men were clinging.

Smashed lifeboats were sinking. I saw something which I
shall never forget. A few hundred yards ahead a whaler was
drifting, with its full crew but the oars with their blades
broken. I steered towards them to offer them a tow and some

encouragement. Going alongside my bow man said, gravely:
'They will not need a tow, Lieutenant.'

They were all dead or dying. The crew of seven were sitting
on their thwarts leaning forward, their hands on the oars,
just as if they were awaiting the order to give way. One man
appeared unhurt until we came close to him and then we saw
that the back of his head had been blown off and the inside
of his skull scooped out, so that it was all white and clean.

A cutter, the biggest of the pulling boats, was going round the
roadstead, picking up the bodies of men, and parts of the bodies
of men, as happened when a passing bow man gaffed a jumper
floating by and discovered that it contained the intestines of
the man to whom it had belonged and nothing more.

The commission of inquiry, which investigated the disaster, reported that there was no trace of sabotage. Evidence showed that
the explosion had been caused, in the first place, by the spontaneous
combustion of Poudre B, as had been the case in the *Iéna*. In the
Liberté it had melted some of the ammunition racks, which had
collapsed, and 735 rounds of 7.6-inch ammunition, together with
4,600 rounds of 65 and 47-millimetre had fallen into the fire already
started, and had exploded.

However, not all the ammunition in the ship at that time had
exploded. Ten years later, shells which had spent a decade at the
bottom of the sea, were removed from the after magazine in the
wreck and loaded on a lighter to be towed out to sea and sunk,
but while the lighter was on its way across the roadstead, the
ammunition suddenly exploded.

The position of the valves for flooding the forward magazine in
the *Liberté* was the subject of much criticism, since the valves were
immediately above the magazine itself, so that if fire did break out
in the magazine it would be practically impossible to approach the
valves to extinguish it.

7

The Guns of Jutland

From the first day of World War I the North Sea had been the focus of the entire naval war, with the British Grand Fleet in the Orkneys watching the German High Seas Fleet behind the Friesian islands. On two occasions, in November and December 1914, German battle-cruisers raided English coastal towns. A third raid was intercepted by the British on 24 January 1915, and the German armoured cruiser *Blücher* was sunk. Thereupon the Kaiser intervened in person in order that the big ships should no longer be risked.

Anticipating an order of this kind, the German naval high command had been casting around to find an alternative to the coastal raids as a way of harming the British at sea, and decided on a campaign of unrestricted submarine warfare, which meant sinking merchant ships without warning.

For this, the first round of the submarine war, neither side was properly prepared, the British being without anti-submarine equipment and experience, while the Germans had only twenty-four U-boats fit for the task with which they were now faced. Nevertheless, the Allies lost a million tons of shipping in the period of the first unrestricted submarine warfare campaign, which lasted from February to September 1915, when attacks on passenger liners ceased, following American protests at the loss of the lives of their citizens in the sinking of the *Lusitania* and *Arabic*. A few weeks later Pohl, the C-in-C of the High Seas Fleet, retired on grounds of ill-health and almost immediately died. His place was taken by Vice-Admiral Reinhard Scheer, who proved to be one of the outstanding sea officers in modern history. His first concern on assuming command was to search for a way in which he could cut down the numerical superiority of the British fleet. He did not believe that it would be possible to bring on an all-out battle in which the Germans would have any chance of success. On the other hand, what he believed he might be able to do would be to lure out a detached part of the Grand Fleet and fall upon it with his whole force. Now that the U-boats were freed from their preoccupation with the unrestricted submarine warfare campaign, Scheer thought that he would be able to station groups of U-boats off British ports (this was before the days of 'wolf packs' in World War II), which could attack the units of the British fleet as they sortied after being lured out to sea by the High Seas Fleet. The original lure had been a raid on Sunderland, similar to the one which Hipper's battle-cruisers had made on Lowestoft the preceding April – undertaken in order to attract British attention from the Easter rising in Dublin.

At the end of May, bad weather and repairs to the battle-cruiser *Seydlitz*, mined during the Lowestoft raid, had caused the postponement of the planned bombardment of Sunderland and the scheme, originally intended to include co-operation with U-boats and Zeppelins, had to be slimmed down. It ended up as a cruise by the High Seas Fleet up the coast of Denmark, where, it was believed, the British had clandestine watchers who would report that the German fleet was at sea.

Scheer, for this operation, had divided his fleet into two parts, the main body comprising 16 modern battleships, 6 obsolete ones, 6 light cruisers and 31 destroyers. Some fifty miles ahead was Hipper with the First Scouting Group of 5 battle-cruisers, escorted by 5 light cruisers and 30 destroyers.

The British fleet was disposed in much the same way; ahead was the Battle-Cruiser Fleet under Beatty, 6 battle-cruisers, with 4 of the most modern and powerful battleships of the *Queen Elizabeth* class, 14 light cruisers and 27 destroyers. Behind them, at a distance of fifty-one miles, was the main fleet under Jellicoe, with 24 modern battleships, 3 additional battle-cruisers, 8 armoured cruisers, 12 light cruisers and 52 destroyers.

Just after 2 pm on 31 May the scouting forces of Beatty and Hipper met – the easternmost of the British light cruisers, *Galatea*, met the westernmost of the German light cruisers, *Elbing*. *Galatea* had spotted smoke to eastward, which was actually coming from a Danish merchant ship letting off steam after she had been stopped by two German destroyers. Professor Potter of the US Naval Academy and Fleet Admiral Nimitz, in their book *Sea Power: Naval History*, commented: 'This contact brought on the battle of Jutland and probably cost the British a decisive victory, for otherwise it is likely that the opposing forces would have met much further North with the Grand Fleet concentrated and with Scheer's escape . . . cut off.'

As it was, Hipper's five battle-cruisers, *Lützow*, his flagship, followed by *Derfflinger*, *Seydlitz*, *Moltke* and *Von der Tann* streaked off to southward towards Scheer, with Beatty's Battle-Cruiser Fleet in hot pursuit. The *Lion* was leading, followed by *Princess Royal*,

Queen Mary, Tiger, Indefatigable and *New Zealand.* Behind them came the four *Queen Elizabeths* of the Fifth Battle Squadron, and before them was a screen of light cruisers.

By 3.55 both battle-cruiser groups were in action, steaming along at top speed. The Germans were moving through patches of mist, while the British were clearly to be seen against the blue horizon. To the naked eye, at a distance of about 16,000 yards, the ships on both sides seemed just dots, but with periscopes and range finders magnifying twenty-four times an amazing amount of detail was visible. Thus the Gunnery Officer of the *Lützow*, Commander Günther Paschen, described after the battle how he could see the turrets of the *Lion*, opposite him in the line, traversing, the guns elevating, firing and then depressing to reload, while the enemy shells coming towards him could easily be seen.

Paschen was firing four-gun salvoes, the two forward 12-inch turrets, followed by the two aft. He attempted one eight-gun salvo but then gave up, as the splash of the shells completely blocked out the target.

At 3.52 *Lützow* scored the first important hit on the *Lion* and it was very nearly the last. There was a flash of red flame alongside *Lion*'s midship turret and a great piece of steel, half the turret roof, flew in the air. A flame higher than the flagship's mast leaped up, burned a while and then died down.

The fire might have destroyed the ship, but this was the celebrated occasion on which the turret commander, Major F. W. H. Harvey, RMLI, having had both legs shot away and being mortally wounded, ordered the magazine doors to be closed and the magazine flooded. For this Harvey received a posthumous VC. The fate from which he had saved his ship, her admiral, officers and crew was to be seen a few minutes later when *Indefatigable* was hit simultaneously on her upper deck by three 11-inch shells from the *Von der Tann*. She turned out of line, apparently sinking by the stern, and was hit again almost at once by another salvo from the *Von der Tann* near her fore turret. There was another explosion, with a cloud of black smoke twice the height of her masts, and when this cleared away she had disappeared.

Her destruction had taken three minutes. By 4.05 she was gone. Out of her complement of 1,019 officers and men, two men only were saved – picked up by a German destroyer.

By mistake, both *Lion* and *Princess Royal* were concentrating on the *Lützow*, instead of each taking a separate target, but they did score three hits on the German ship, the first serious one bursting between A and B turrets, in the forward dressing station where all within were killed.

At 4.26 *Queen Mary* was hit; there was a vivid red flame forward, another explosion amidships, her funnels and masts collapsed inward, a column of smoke more than 2,000 feet high mounted and she turned over to port, down by the bow. Her stern rose out of the water, with her screws still turning and vast quantities of paper blowing out of the after hatch.

It was at this moment, with his own ship still on fire and seemingly about to blow up, that Beatty made his contribution to immortal quotations in the English language when he turned to Chatfield, his flag captain, and said: 'There seems to be something wrong with our bloody ships today.'

He then ordered course to be altered towards the enemy. Of the *Queen Mary*'s crew of 1,294, twenty were saved including one midshipman who made his way out of the sinking wreck to the comparative safety of the sea by running down the side of the collapsed funnel.

For the Germans, nightfall and ever-present smoke now began to trouble their shooting; this was frustrating for Paschen who was conscious that he had been doing well. He had been encouraged early in the action by the sight of two British shells which had fallen in the water alongside the *Lützow*, richocheted off the surface, sailed slowly over the *Lützow*'s forecastle and fallen into the sea on the other bow, travelling so slowly that Paschen could see that they were painted white. These, he knew, were common shell containing the old-fashioned black powder.

Up to now *Lützow* had been firing only HE (high explosive) shell, which Paschen was afterwards greatly to regret for, he claimed later, had he used AP (armour piercing) shell the *Lion* would never have survived. For the time being, however, Paschen's greatest worry was the enormous quantity of water from enemy shell splashes which poured into his control position, the canvas which covered the gap through which the fire control periscope could be moved having blown loose. One seaman spent the entire action sitting on the roof of the conning tower, in the open, wiping the periscope lens but that, of course, did not prevent everyone in the control position being soaked to the skin and staying soaked throughout the battle.

The fighting during this phase of the battle, between 4.15 and 4.43, was described by Beatty as being of 'a very fierce and resolute character'. At 4.30 the German battle-cruisers sighted their main fleet and eighteen minutes later the British saw them as well. Both groups of battle-cruisers then turned around, heading back to the North; Hipper had been leading Beatty down to the German battle fleet. Now the situation was reversed and Beatty was luring the whole High Seas Fleet north to confront the superior power of Jellicoe's Grand Fleet.

Firing between the battle-cruisers slowed down. Hipper reduced his speed to seventeen knots and the British withdrew to the limit of their range as visibility began to diminish. At 5.41, however, the Grand Fleet was sighted by Beatty who turned towards the enemy and opened a heavy fire. Within a few moments *Lützow* was hit four times. Two shells penetrated the deck over the forward 5.9-inch casemates and, bursting, destroyed both wireless offices causing heavy loss of life. A third shell exploded on the middle deck, without penetrating the main armoured deck, but the explosion, taking place just above the Gunnery Transmitting Station, caused fuses to burn out and filled the compartment with smoke.

Paschen wrote later in the *Marine Rundschau*:

> Fire control for a short time broke down, our own smoke was following us and I had to shift control to the after control position. For the moment the enemy had the upper hand and I could not shake off the feeling that things were going wrong.
>
> Then came something quite unexpected. From right to left there came in sight through the periscope a ship, improbably large and close. I saw at once that it was an old English armoured cruiser and I gave the required orders.
>
> Somebody grabbed my arm:
>
> 'Don't shoot, it's the *Rostock*.'
>
> But I could see turrets on forecastle and quarterdeck and I opened fire at about 7,000 yards.

Five salvoes followed rapidly; three straddled the target and Paschen could see a ship blowing up, clearly visible to both fleets.

This was the end of the *Defence*, lost with all on board. The *Warrior* was following and the *Lützow* left her to the *Derfflinger*, her next astern, who quickly disabled her so that she sank the next day. *Lützow*'s attention was now entirely taken up in looking after the enemy battle-cruisers. They were now on her port quarter, about 13,000 yards away, and barely visible. Suddenly *Lützow* was hit forward and aft by shells coming from her port beam. There was nothing to be seen of the enemy except for red flashes. Her turrets were trained on their extreme bearing to port and were firing as fast as possible at the British battle-cruisers. From their direction came a shell which went through the upper deck, abeam of the forward funnel, through the casemate and burst on the base

of B turret, blowing out both doors leading from the casemate onto
the foredeck. The explosion took place right under the bridge and
the conning tower without, however, damaging the bridge.

'A' turret reported that the right-hand gun was out of action and
that the magazine was gradually filling with water. *Lützow* was now
on a southerly course when suddenly, four points on the port
quarter, a British battle-cruiser of the *Invincible* class, clear and
comparatively near, was coming out of the mist.

Paschen trained his guns on the newcomer. In fifteen seconds his
three remaining turrets were in action and it was only a few
moments later that a great flame and much smoke seemed to cover
the forward half of the enemy. Then the hull was rent apart and
the two halves of the ship stood upright on the sea bottom and
remained there for some hours before they finally sank. After the
war British divers went down to the wreck in order to establish her
exact position, for it was from this spot that many ships calculated
their position during the battle. As for *Lützow* herself, B turret was
jammed hard aport with smoke coming out of all openings and
the turret crew giving no answer to attempts to communicate with
them.

This was the beginning of the end of the ship. At 6.43 she left the
line and slowed down, no longer able to hold at bay the water
forcing its way into the damaged bow. Four German destroyers
made an almost impenetrable smoke screen between the *Lützow*
and the enemy. For the first time in 3½ hours Paschen left the fire
control position; as he emerged Hipper was transferring his flag.
Waiting for his destroyer, *G 39*, to come alongside he was relaxed,
talking to the officers and men gathered around, thanking them and
praising them for their work. The other four German battle-cruisers
steamed by, Paschen noting that both *Derfflinger* and *Seydlitz* lay
deeper in the water than usual.

More red flashes came out of the twilight. Communications between Paschen and his guns had been completely wrecked by enemy
hits. 'If only we had had the director,' wrote Paschen afterwards,
referring to the British device with which all but two of the battleships of the Grand Fleet at Jutland were fitted and which enabled
them to aim and fire their guns from one central point.

Two or three enemy shells shook the *Lützow* aft. One burst in
the 'tween-decks between C and D turrets in the after dressing
station, killing many amongst the wounded, and the sick berth
attendants. The same shell cut the power cable which supplied D
turret and which in this part of the ship ran above the armoured
deck instead of below it. D turret went over to hand working, which
meant that it was all but out of action. In addition, this shell had

penetrated the armoured deck above the magazine, without causing the magazine to explode.

At this moment the whole British battle line began to steam past, but the visibility was such that the German ship was unseen, and her people began to do what they could to carry out repairs.

The surviving members of the gun crews of B turret, who had been spared because of a bulkhead between the two guns, had got their gun back into action so that the ship had four out of eight of its 12-inch guns ready, as well as three out of the fourteen 5.9-inch guns of the secondary armament. Communications between parts of the ship were disrupted and it was necessary to organise chains of messengers to pass orders. *Lützow* was now so down by the head that her foredeck was awash. The whole forepart of the ship was full of water and a number of men were cut off in a compartment on the upper platform deck. There was no way in which they could be helped until the ship was drydocked and the prospect of that was a long way away and, indeed, was never to be possible. The magazine of A turret was full of water and that of B turret was flooding but, for the time being, this could be kept in check.

At about this time the ship went to night action stations. The gunnery spotting officer came down from the control top. He was wearing the brim of his cap, the cover having been torn off. The ship was now making five knots and sea was breaking over the forecastle and splashing around the piles of empty shell cases. The Executive Officer reported that there was 7,500 tons of water in the ship and that they would not be able to keep afloat after eight o'clock next morning, at the latest. They attempted to steer stern first but the ship would not answer to her helm.

Shortly before three o'clock the order was given, 'All hands aft'. The four destroyers which had stood by all night came alongside and preparations to abandon ship began.

The forecastle was now six feet under water. The sea was streaming through the doors leading to the battery amidships and pouring through shell holes down into the middle deck. Paschen and others went through the ship, looking for wounded men or men who had simply fallen asleep after the strain of the past twenty-four hours, and he did find one fast asleep on a work-bench.

They then abandoned ship; there was nothing that could be done for the men cut off on the platform deck. The others, including all the wounded, were transferred to the destroyers. It was now growing light. A turret had disappeared beneath the water, B turret above it stood up like an island. At the bridge the water was washing over the upper deck and the stern was six feet out of the water so that one torpedo from a destroyer was enough to finish her off.

Lützow and her sister-ships *Derfflinger* and *Hindenburg* were generally judged, by British and Germans alike, to be the best mixes of gunpower, speed and protection ever produced, except for the fact that the Germans in World War I had not felt that they had sufficient reserves of oil fuel to rely upon oil for the propulsion of their big ships. Accordingly, while the British had gone over to oil-fired capital ships in 1912, the Germans never did so until the foundation of Hitler's navy.

However, there was one mistake in the otherwise exemplary design. Inside these ships, parallel to the ship's side itself, was a thinner internal armour belt which stood the big German ships in good stead except, in this case, to save space the belt had been omitted behind the submerged torpedo tubes and it was at this very spot that the *Lützow* was hit by two British shells.

8

The Battle of the Adriatic
Szent Istvan and *Viribus Unitis*, 1918

Early in 1910 a visit was paid by the heir to the Austro-Hungarian Empire, the Archduke Franz Ferdinand, to the head of the Vienna branch of the House of Rothschild which startled those who were professional or amateur watchers of the Imperial Court, of the Vienna Bourse and of the naval position of the powers in the Mediterranean.

To stress to the public the purport of his visit the Archduke had put on the full-dress uniform of an admiral in the Austro-Hungarian navy, to which he was entitled although he had never served at sea in his life. This full-dress uniform was mildly surprising, but what really attracted comment was the fact that the heir to the throne had condescended to call on a banker, and a Jewish banker at that. However, it was known that the Archduke was seeking a loan – a big loan – to expedite the laying down of four dreadnought battleships needed to secure for Austria a place, albeit the lowest, among the seven principal naval powers of the world.

The official reason given for this expensive step was that it would increase Austria's value as an ally to the other two members of the Triple Alliance, Germany and Italy. But the true motive was the feeling that Austria's navy should more or less keep pace in size with that of Italy, which was soon to number six dreadnoughts. It was therefore believed in Vienna that at least four Austrian dreadnoughts were needed. The number of dreadnoughts being the yardstick by which naval strength was measured amongst the 'Top Eight' naval powers (Britain, Germany, the United States, France, Japan, Russia, Italy and Austria), while lesser powers believed that even a single dreadnought might be essential to their security. This accounted for minor naval building races between Argentina, Brazil and Chile and, on the other side of the globe, between Greece and Turkey. Spain, in competition with no one, built three small dreadnoughts.

There was an additional reason why Austria should make haste to build the new ships: the principal Austrian yard capable of that work, the Stablimento Technico at Trieste, was in poor straits financially and about to dismiss a great proportion of their skilled hands for lack of work. Once this happened it would be impossible for them to reconstitute their workforce. There was a further snag; the Financial Delegations of Austria and Hungary, responsible for voting the Budget of the complex of peoples which made up the Empire, would not meet until the following year. A hasty meeting between the Directors of the Stablimento and Admiral Montecuccoli, who despite his Italian name was Kommandant of the Austro-Hungarian navy, tried to find the money needed for an immediate start to the work and it was then that someone produced the idea that the Rothschilds should be asked for a bridging loan. Hence Franz Ferdinand's appearance, in full-dress uniform, seeking money which the House of Rothschild agreed to provide.

There was just one more obstacle to be overcome before the programme could begin. The Hungarian part of the Delegations refused to provide their share of the money unless one of the four projected ships was built in Hungary's only yard, the Danubius at Fiume, which had neither the facilities nor the experience necessary for the task. The expansion of the yard required a great deal of money which increased the cost of the Danubius ship (*Szent Istvan*) and, as will be seen, may have contributed to her loss.

The lead ship of the four units of this class, the *Viribus Unitis*, spent all the six years of her life as flagship of the Austrian fleet, and it fell to her to bring back the bodies of Franz Ferdinand and his wife to Pola after their murder in June 1914.

The war in the Adriatic settled down to a somewhat dreary pattern resembling that of the struggle in the North Sea, with the Austrians based on Pola controlling the northern waters of the Adriatic and the convoy routes down the Dalmatian coast for the supply of the Austrian troops in the Balkans. The Allies, Italian, French and British, and later the Americans, were based on Brindisi controlling the southern waters and protecting the barrier of nets and mines intended to prevent the escape of German and Austrian submarines into the Mediterranean.

The war developed in a strange manner on account of this mixture of races, languages and loyalties. Lower deck personnel divided as follows, by race: 34.1 per cent were South Slavs (31.3 per cent Croatian, 2.8 per cent Slovenes), 20.4 per cent Magyars, 16.3 per cent German Austrians, 14.4 per cent Italians, 11.8 per cent Czechs, Slovaks and Ruthenians and 3 per cent Poles and Rumanians. German Austrians and Czechs, as a rule the more sophisticated of the races of the Empire, were mostly engine room personnel and electrical specialists, as well as gun crews for the heavy guns of the big ships, signalmen and coxswains. Secondary armaments were usually manned by Hungarians, while the Croatians proved best at boat work, as stokers and, in general, the tasks aboard ship which called for exceptional physical strength and endurance.

Dealing with this mixture of races and languages required special qualities from their officers, all of whom had to be able to express themselves clearly in four languages. Ratings had to understand orders given in German and to speak fairly well the two commonest of the languages in the navy, Croatian and Italian. As examples of what this meant in practice, there was a master mariner of Italian race and Austrian nationality who served as a pilot for various commando-style raids on the Austrian coast. Captured by the Austrians, he was executed. There was another Italian of Austrian nationality who ran an organisation for the sabotage of Italian warships, which accounted for two Italian battleships, the *Leonardo da Vinci* of 22,000 tons, and the *Benedetto Brin*, of 13,215 tons. It was claimed that the destruction of these and other ships was caused by the introduction of bombs disguised as lumps of coal. The leader of this ring of saboteurs was taken prisoner by the Italians while leading a commando raid, but as he was in uniform at the time the Italians, in accordance with the laws of war, treated him as an ordinary prisoner of war and, when the war was over, released him.

A number of sabotage attempts were also made against Italian railway junctions, factories and various military installations from a base which was in an annexe of the Austrian Consulate-General in Zürich, but in those days of innocence in espionage the headquarters of the organisation was left unguarded at night. Reliance was placed by the Austrians merely on the protection which diplomatic usage gave to foreign missions. The Italians, however, broke into the offices of the Consulate-General and removed the safe, a gigantic construction of steel with double walls, the space between them filled with tear gas.

Just as the Allied navies tried to prevent German submarines leaving the North Sea via the Straits of Dover, attempts were made to seal the exit from the Adriatic into the Mediterranean. By the middle of 1918 a barrier some forty miles long had been stretched across the Straits of Otranto, consisting of a steel net 150 feet deep, hanging from 428 buoys stretched 30 feet below the surface, with 1,200 mines attached. Patrolling the nets were Italian, British, US and French ships, together with observation balloons and aircraft, a total of 20 submarines, 40 destroyers, 8 sloops, 36 MAS (Motoscafi Anti-Sommergibile), a flotilla of American 'sub-chasers', 48 trawlers and 76 drifters. Of these, between half and a third were on station

at any one time. The nets, mines and patrol craft were a permanent
target for the Austrians and altogether they attempted no fewer
than seventeen raids, the most successful of which was that on
15 May 1917 when they sank fourteen British drifters. The Austrian
force of light cruisers and destroyers on this occasion was commanded
by Captain Nicholas von Horthy, afterwards Regent of Hungary, in
the light cruiser *Novara*. The force was intercepted on its way back
to Cattaro (now Kotor); the *Novara* was badly damaged and had to
be towed home, while Horthy was seriously wounded. The combined
British and Italian intercepting force was finally driven off by the
Austrian armoured cruiser *Sankt Georg*. In the following year
Horthy was appointed C-in-C of the fleet, over the heads of forty-
eight captains and admirals.

The Austro-German victory at Caporetto, the battle which began
on 24 October 1917, completely changed the situation at the
northern end of the Adriatic. Ever since the outbreak of war in
May 1915 the Front in Venetia had been almost completely static.
Now the Austrians began an advance along the coast which brought
them to within twenty miles of Venice. To support this advance the
Austrians used two of their oldest and smallest battleships, *Wien*
and *Budapest,* of 5,650 tons, to bombard the Italian rear areas along
the coast.

To stop this the Italians decided to use a pair of their MAS Anti-
submarine Motor Boats, corresponding to the British CMBs (Coastal
Motor Boats). Two boats, MAS *9* and *13*, were towed by torpedo-
boats close to the spot in the Muggia channel, outside Trieste, where
the *Wien* and *Budapest* were known to lie up at night after a day's
bombardment work. The night of 9/10 December was dark and
foggy. The two boats cast off at 22.45 and entered the channel and
Lieutenant-Commander Rizzo, a former officer of the Italian merchant service, leading the expedition, then ran his craft alongside
the mole at the entry to the harbour, climbed onto it and coolly
walked along checking on the positions and routine of the sentries.
Back on board he soon came up against the first of the obstructions
placed by the enemy which was a double 3-inch cable secured to
buoys. When this had been severed the Italians encountered a 4-inch
cable under water, and finally they came to five cables hung in
festoons – presumably designed to ensnare the screws of passing
ships. About two hours later they had finished cutting the cables
undisturbed, though they could hear voices ashore and see a light
in a nearby cottage.

Then, running dead slow on their electric motors, the MAS
entered the gap in the boom which had been made. After twenty
minutes a dark mass appeared to port, and Rizzo told MAS *13* to

stay behind, with orders to fire her torpedoes as soon as any noise was heard.

At 0232 there were two explosions and two tall spouts of water. Immediately afterwards there were two more explosions as MAS *13* fired, missing *Budapest* but hitting the embankment while both boats, the throttles of their 350 hp Isotta-Fraschini motors wide open, roared off into the night, leaving the *Wien* sinking and taking with her forty-six members of the crew, the shouts of the drowning men mingling with the joyous cries of *'Evviva il Re!'* from the Italians.

The success of the Italian raid made a great impression on the Austrians who were interested in small, fast torpedo craft and who had already designed and built the world's first hovercraft as a motor torpedo-boat or anti-submarine craft. This vessel, however, had been unsuccessful and, after further experiments, they resolved to capture an MAS to discover the details of its construction. Accordingly, on the night of 4 April 1918, a commando-type operation under Linienschiffsleutnant (Lieutenant-Commander) Veith, with five officers and sixty men landed near Ancona and entered the town, with the idea of cutting out one of the two MAS which were reported as lying there, and carrying it off to an Austrian port.

On their way to the harbour they were stopped several times but disarmed Italian suspicions by the explanation that they were British. In Ancona they discovered that the MAS were not where they were supposed to be. They then boarded a fishing boat in which they hoped to escape, but could not get the engine to start and soon found themselves surrounded by a large party of Italian troops who had been alerted by one of the Austrians, a man of Italian origin. They were made prisoners of war.

'When the United States Assistant Secretary of the Navy, Franklin D. Roosevelt, visiting Italy, inquired why the Italians did not even put to sea for training exercises, the response was that it was superfluous because the Austrians also refrained from such activity. Roosevelt noted in his diary: "This is a naval classic, which it is hard to beat, but which perhaps should not be publicly repeated for a generation or two." ' [12]

When Horthy was appointed to command the fleet on 1 March 1918, he decided to carry out a raid on the Otranto barrage. Not many people knew then, and no one knows now, why he did this. In theory an attack on the barrage would free the Straits for the exits and entrances of the U-boats, but at this time the barrage was not proving much of an obstacle and the proposed operation did not seem worth the risks entailed.

It was intended that the raid should be carried out by two light cruisers *Novara* and *Helgoland*, with four destroyers, while two more light cruisers *Admiral Spaun* and *Saida*, with four torpedo-boats and eight aircraft, were to attack the Italian seaplane base at Otranto itself. Four dreadnought battleships (*Viribus Unitis*, *Tegetthoff*, *Prinz Eugen* and *Szent Istvan*, 20,000 tons, with twelve 12-inch guns) and three pre-dreadnoughts (*Erzherzog Ferdinand Max*, *Erzherzog Friedrich* and *Erzherzog Karl*, 10,600 tons, four 9.4 inch guns) were to act in support. Nevertheless, the concentration of Allied strength seemed certain to be able to withstand the attacking Austrian light forces, while the barrage itself was not, despite the forces which had been concentrated for its defence, effective at this time. According to Herr Karl von Lukas in *Marine Gestern und Heute* during the first six months of 1918 neither German nor Austrian U-boats had been lost making the passage in or out of the Adriatic.

Herr von Lukas, therefore, dismisses the idea that the raid on the barrage was the real purpose of Horthy's planned operation. According to him the four years of idleness, which the big ships had spent since the war, had begun to have an effect similar to that experienced in the German High Seas Fleet where mutinies had taken place during the summer of 1917. In the Austrian navy a torpedo-boat had gone over to the Italians on 5 October 1917 and in the following February there had been mutinies among the crews of the cruisers based on Cattaro, which had ended with the execution of four of the mutineers. These events, says Herr von Lukas, had convinced Horthy that the fleet should take part in an important operation, which would have a far better effect on the morale of the crews than the death sentences and executions.

Horthy's plan was to send his fast light cruisers and destroyers down the Adriatic to do what damage they could to the barrage. This, he believed, would lead the Italians and British light craft to appear and try to cut off their Austrian opposite numbers on their way home. These Allied ships would then be exposed to the Austrian battleships, which would have arrived secretly on the scene having left Pola two nights previously.

A feature of the Austrian plan was that their battleships should operate singly with small escorts of destroyers and torpedo-boats, so that, in whatever direction the Italians emerged from their base at Brindisi, one or more of the seven Austrian heavy ships would be waiting for them. Of these battleships *Szent Istvan* and *Tegetthoff* would leave Pola on 9 June, while their sister-ships, *Viribus Unitis* flying Horthy's flag and *Prinz Eugen*, had sailed the day before for Cattaro.

As soon as *Szent Istvan* and *Tegetthoff* cast off to leave harbour things began to go wrong. First, the gate in the boom closing the harbour was not opened on time, and the two ships were delayed for an hour. Then *Szent Istvan* ran a hot bearing, so that the speed of the two ships had to be reduced from 17.5 knots to 12.5 knots. The cumulative effect of these two delays was matched by another which was to seal the fate of the ship. Two MAS, *15*, commanded by Rizzo who had sunk the *Wien*, and *21* were patrolling and sweeping for mines between the islands of Premuda and Lussin. The patrol went on slowly, waiting for the moment to turn back which, in theory, should come in time to allow the MAS to get well clear of the enemy coast before dawn. This was necessary because during the patrol the Italians were so close inshore that they could smell the scent of the pine trees on the Austrian islands. Then Rizzo noticed that his watch had stopped. A quick check and he saw that he was twenty minutes late at the turn, an unpleasant feeling, followed by sighting a smudge of smoke which soon grew up into a heavy cloud.

In passing it may be noticed that the stopping of Rizzo's watch is described in the Italian accounts as 'a small technical fault'.

The two MAS headed for the smoke. There was enough light for the Italians to make out six torpedo-boats; Rizzo reacted to this by ordering both MAS to attack. As they drew nearer to the enemy they saw that the torpedo-boats were not simply on patrol but were escorting two big ships. The MAS to the eastward were against a sky which would be paling within half an hour, so time was short, but there was just enough for the Italians to circle round the big ships and attack them from westward where the sky was still dark, while the escorts would begin to show up against the dawn. Both the big ships were battleships, comprising half the enemy fleet of first-class warships.

To get close enough not to miss, provided that the torpedoes ran properly, it was necessary to pass between two of the escorting torpedo-boats, now only fifty yards away. Rizzo achieved this without being spotted and fired both his torpedoes at 350 yards. A splash and the torpedoes disappeared. Then two patches of bubbles suddenly stretched themselves into two milky-white tracks, tearing towards the enemy. Sixteen seconds to go. Rizzo had ordered full speed as soon as he fired, and his engines were making so much noise that he could not hear the explosion but, suddenly, there were two satisfying columns of water and smoke amidships. There was no sign of anything untoward with the second battleship. Clearly MAS *21* had missed. Neither of the MAS carried reload torpedoes, so that there was nothing for them to do but streak away to the dark horizon and disappear.

When dawn came there was a sight which has become familiar to thousands of cinema goers all over the world. There have been very few films of modern naval war which have not included shots of the sinking of the *Szent Istvan*, first with a slight list to port, with her crew falling in on deck. The event was filmed from the *Tegetthoff*, which was manoeuvring cautiously to take her in tow. Meanwhile, as the *Szent Istvan* slowly listed further to port, both boiler rooms began to fill. The pressure of water caused the steel bulkheads to bulge and then, as the pressure continued, the rivets one by one at first and then in volleys broke up, and the rivet heads shot across the compartments making a noise like pistol shots in the confined, metal-walled spaces. Electric light bulbs burst and the stokeholds were in darkness, save for the glow when a boiler door was opened. To make matters more difficult, almost all the crew of the *Szent Istvan* were new to the ship, having joined only a month previously. It was not surprising, therefore, that with the water rising from their ankles to their knees there was a sudden rush from the boilers to the ladders leading to what might prove to be safety, but officers and petty officers checked the rush. If the fires died the ship died too, and so did her crew. Everything depended upon steam being kept up; ventilation, engines, electricity for lights and for signals.

The coal bunkers were flooding as well, and as they were already nearly full of coal the air pressure increased and blew off the heavy steel lids on the coal chutes which led from the lower deck to the bunkers. The lids came off with a series of great crashes and columns of water, like geysers, splashed up to the deck head and then came crashing down.

The lights flickered throughout the ship and then went out. In darkness, lit only by hand lamps, men in the engine and boiler rooms stuck to their posts until there was no more steam and the list on the ship became so great that a man could no longer stand upright. Then they made their way on deck.

Tegetthoff and the escorts were gathered around. *Szent Istvan* lay over on her side so that her masts and funnels were flat on the water. Some of her crew were making their way down the ship's side, climbing over the bilge keel, picking their way through the marine growth on her green bottom, for she had never been drydocked since completion. Others were swimming, and others had already been picked up by boats from the escorts. Then *Szent Istvan* rolled right over, water spurting from gaps in her sides, amidst screams and whistlings of escaping air and the crash of tearing steel. Suddenly the hull, now completely upside down, rose four or five feet, lightened because her four big gun turrets had torn themselves from their mountings. This was her last struggle. Her bow dipped below the sea and she slipped down, taking with her eighty-nine officers and men.

On receiving news of this disaster Horthy cancelled the planned operation, because once the Allies knew that the Austrian big ships were at sea they would not take the bait offered by the raiding small craft.

The sortie of 9/10 June, which ended in the loss of the *Szent Istvan*, was the last operation which involved the Austrian navy outside coastal waters and on 15 June the Austrian army began its last offensive. At the end of a week this was clearly a failure and four months later, on 24 October, the Italian counter-offensive began, the intervening time having been spent gathering a concentration of strength which would prove overwhelming. Italy was very near exhaustion and General Diaz, the Italian Commander-in-Chief, knew that he was preparing his country's last throw. While he was making ready for this the strength of the Central Powers was rapidly ebbing away – Germany's in the last great battle of the Western Front, and Austria's in the strains of the separate nationalisms of which it was composed, fanned by very successful propaganda campaigns directed by groups of exiles and emigrants in Britain, France and the United States. By the autumn of 1918 it was clear that the end of the war was at hand and big powers and small powers, masses and minorities were staking the claims which they would put before the peace conference. One of the first groups to submit its claim was that comprising the Croats and Slovenes who, together with the Serbs, make up what is now Yugoslavia.

A delegation from the Yugoslav nationalities was collected and sent on its way to Paris to seek official recognition and, incidentally, to offer to the Allied cause the Austrian navy which they rightly believed would come to them. This delegation was detained for several days by the Italians, who were resolved to have no rival in the Adriatic. By the time the delegates were allowed to proceed things had altered considerably in Italy's favour, so that the final operation of the naval war in the Adriatic could be launched by the Chief of the Italian Naval Staff, Admiral Thaon di Revel, who signalled the Admiral Commanding Venice: 'The Navy must, before the final acceptance of the Armistice, contribute to the defeat of the enemy.'

This contribution was a follow-up of Rizzo's successful attack on the *Wien* outside Trieste, after which the Italians had developed two new methods of assaults on harbours protected by guns and searchlights, as well as by booms and nets. First was a boat called *Grillo* (Grasshopper) which was a kind of sea-going tank, with caterpillar tracks which enabled it to climb over obstacles. One of these boats

got into Pola on the night of 13/14 May, climbing over the first
barrier but was then spotted and destroyed from the shore.

Something even more inconspicuous was needed and it was de-
vised by two Italian officers, Major Rossetti of the Corps of Naval
Constructors, and Surgeon-Lieutenant Paolucci, a naval doctor who
was presumably covered by the Geneva Red Cross Convention, but
who never appears to have been criticised for his remarkable efforts
as a designer and then user of a very successful infernal machine.

This craft, called *Mignatta*, began life as a German torpedo which
had failed to explode. Rossetti got to work; in the after end he
installed a small noiseless motor driven by compressed air which
was contained in the midships section of the torpedo, at the forward
end of which were two metal containers, each holding 440 pounds
of trinitrotoluolammonal and an adjustable time fuse. On the night
of 31 October, towed by a torpedo-boat and escorted by another
torpedo-boat and two MAS, the little force sailed under Com-
mandante Costanzo Ciano, father of Mussolini's Foreign Minister.

Both Rossetti and Paolucci were very experienced swimmers and
they needed to be, since the *Mignatta* had to be accompanied by its
two-man crew swimming alongside it, and not astride as on the
Wellman craft of World War II. They were accustomed to night-
long swims in the Venetian lagoons but discovered, with something
of a shock, that the water off Pola was much colder than that off
Venice.

Wearing the headgear of their wet suits, which were shaped like
bottles so that their appearance drifting by would cause no comments,
they reached the breakwater and manhandled the *Mignatta* around
the gap at the entrance to the harbour. Paolucci swam along the
mole to determine the best way into the harbour. At the very feet
of a sentry he was overcome by a desire to cough. He coughed and
nothing happened. He continued his reconnaissance and found the
gap he sought, but the current was now coming out of the harbour
and it began to carry the *Mignatta* out to sea, so that Rossetti was
obliged to switch on the craft's motor and head it into the centre of
the port.

Paolucci dived under the next net, pulling the *Mignatta* after him
while Rossetti, bracing his feet against the net, shoved and steered
the craft from astern. They went on to pass seven barriers and then
discovered that their compressed air was running out. However, they
decided to ignore this danger and pushed past the darkened battleships of the *Radetsky* class and then found three more battleships,
illuminated from stem to stern.

At this moment the *Mignatta,* without warning, began to sink.
Desperately Rossetti checked his controls and found that he had

forgotten to close the valve which regulated the flow of water into the *Mignatta*'s hull as he strove to get her proper trim. He succeeded and the *Mignatta* slowly advanced against the harbour and came alongside the battleship *Viribus Unitis*, the Austrian flagship, at 0420. The current carried them away and then, painfully, they crept back again. Rossetti disconnected the mine from the body of the *Mignatta*. He swam back with it to the enemy ship and tried, with frozen fingers, to set the time switch. In the end he succeeded just as a searchlight beam, sweeping the surface of the water, showed up the two swimmers. A boat put out from the ship and the two Italians were hauled on board and taken back to the battleship they had just arranged to blow up.

As they came up the ship's side they saw that the old Austro-Hungarian cap ribbons reading *K-u-K Kriegsmarine* (Imperial and Royal Navy) had been replaced by strips of paper bearing the single word 'Jugoslavia'. The Italians now learned that on the previous day the Austrian Emperor, his empire breaking up into its component racial parts, had bestowed his navy upon the Yugoslav National Committee. Horthy had hauled down his flag that day, 31 October, and gone over the side of the *Viribus Unitis* with his flag on his arm, while the senior Yugoslav officer present, Captain Vukovic, became Commander-in-Chief of the fleet.

On the same evening the red-white-red Austro-Hungarian flag had been hauled down for the last time at Pola; in theory it was to be replaced by the blue-white-red Yugoslav flag, but very few of these actually existed and, for that reason, many of the ex-Austrian ships hoisted the Montenegrin flag.

Faced with these facts Rossetti and Paolucci made up their minds to tell Vukovic what they had done and he gave the order to abandon ship. The Italians were overlooked and saved themselves by jumping overboard, only to be picked up again and brought back to the doomed ship.

It was daylight now and at 6.30 am the mine exploded. A comparatively small column of water about ten feet high, almost a splash, rose in the air and the ship began to lie over to starboard. The vicissitudes which the crew had undergone in the past few days were reflected in the ship now. According to the Austrian official historian, the former Linienschiffsleutnant Hans Sokol, damage-control procedures collapsed completely. In the middle of the confusion Vukovic pointed out to the two Italian prisoners a rope hanging over the side, down which they slid back into the water where they were picked up for the third time and spent the last days of the war and the first days of the peace on board the training ship *Hapsburg*.

Within fifteen minutes the *Viribus Unitis* capsized and the last that was ever seen of Vukovic, admiral for a day, was of him standing alone on the upturned hull of his flagship as she slipped beneath the surface.

At sunrise on that same morning the former Austrian warships at Cattaro and Sebenico hoisted the Austro-Hungarian colours for the last time, fired a salute of twenty-one guns, hauled down the Austrian flags and replaced them with those of Yugoslavia. On this day, when the Austrian navy officially ceased to exist, the authorities in Vienna issued a list of promotions of senior officers.

It was long suspected that the Italians launched the *Mignatta* attack knowing that the Austrians had already agreed to surrender, the acceptance of which was delayed by the Italians for considerations of prestige and a desire to destroy any possibility of Yugoslavia becoming an important power. In this, as far as the ex-Austrian navy was concerned the Italians were successful, for in the end the Yugoslavs were left with only a few small torpedo-boats. On the other hand, Italian attempts to take virtual control of the Dalmatian littoral failed after arguments which dragged on until January 1924.

The strong probability that the Italians knew that the Austrian fleet had surrendered before the *Mignatta* attack was ordered, finally came to light after an interval of sixty years. In the last days of the Austrian Empire the different nationalities which had made it up maintained semi-official liaison between each other. Thus there was at this time in Pola a delegation from the Czech workers organisation which reported to Prague, and it was in the archives there that a report was found in 1978 giving details of relations between Austrians, Italians and Yugoslavs, published in the *Marine Rundschau* of July 1979 by Herr René Greger.

9

Fifteen Battleships Sunk in an Afternoon by Mistake

Scapa Flow, 1919

When the Allies were drawing up their conditions for an Armistice in October and November 1918 the fundamental point was that once it was signed the Germans must not be able to resume hostilities, for the British and French armies were very close to exhaustion. To ensure that Germany would be in no condition to restart the war, the Allies determined to secure the withdrawal of the German army thirty miles behind the Rhine, and the surrender of a formidable quantity of war material: 5,000 guns, 25,000 machine-guns, 3,000 trench mortars and 1,700 aircraft, together with motor transport, locomotives and rolling stock.

Decision by the Allies on these points was reached easily enough, but there was disagreement over the future of the German navy. The British, who had borne the brunt of the war at sea, wished to secure the surrender of the 16 most powerful capital ships, battleships and battle-cruisers, 8 of the most modern light cruisers, 50 destroyers and all existing submarines, of which there turned out to be 150. Here Marshal Foch, the Allied Commander-in-Chief on the Western Front, intervened. Provided that Germany was rendered powerless on land he did not believe that she could be pressed too hard. The German surface fleet, he said, had had no influence on the war and it would therefore be sufficient if, during the period of the Armistice, it was concentrated in the Baltic.

Admiral Sir Rosslyn Wemyss, the British First Sea Lord, protested firmly, pointing out the immense proportion of the British war effort that had been needed to keep the German High Seas Fleet in check, an effort which would have to be maintained if the Germans were allowed to retain control of their ships. Finally it was decided that the principal German ships, those whose surrender had been originally demanded, should be disarmed and interned in an Allied or neutral port.

While these discussions were going on, the High Seas Fleet had ceased to exist as an organised force. In a final bid to regain the prestige which it had lost by the passive role forced on it by its numerical inferiority to the British Grand Fleet, and also in the hope that a serious blow might be dealt to the flow of supplies from the United Kingdom to the Front in France it was decided to send a force of light cruisers and destroyers to raid the mouth of the Thames and the coastal flank of the Allied armies in France and Belgium. In support, a few miles to the northward, would be the big German battle-cruisers at Black Deep near the Sunk lightship, eighteen miles north of Margate and twenty-six miles east of Foulness Point. At the same time the battleships of the High Seas Fleet would come down as far south as 51° 40′, that is, a line between Clacton-on-Sea and the Oester Scheldt. After the light cruisers and destroyers had carried out their raids, the whole German fleet would concentrate and steam northward, expecting that on the evening of 31 October they would meet the Grand Fleet steaming south, when the final naval battle of the war would be fought.

On 26 October rumours of the impending operation began to spread around Wilhelmshaven, started by a young officer at a party. Next day the High Seas Fleet began their preliminary moves; the U-boats involved had already taken up their positions, having been withdrawn from the patrol areas which they had been occupying in their campaign against merchant shipping. This had been called off at the insistence of President Wilson, as a preliminary to an Armistice.

Almost immediately things started to go wrong for the Germans. In a number of ships, beginning with the minelaying cruiser *Strassburg* ordered to load mines at Cuxhaven, members of the crew vanished over the side as their ship was leaving Wilhelmshaven and had to be rounded up. *Regensburg*, the senior officer's ship of the Fourth Scouting Group flying the broad pennant of Commodore Karpf, had similar trouble and when the battle-cruisers began to undock several hundred men from the *Derfflinger* and *Von der Tann* jumped ashore and got themselves lost in the confusion of the dockyard, and they too had to be rounded up and brought back to their ships; this was, however, done quite peaceably.

The rumours going through Wilhelmshaven, together with the concentration of the fleet in Schillig Roads, between Wilhelmshaven and the sea, convinced everyone that a big operation was planned and, on the night of 29 October, the crews of the battleships *Thüringen* and *Helgoland* were on the edge of mutiny. The mood spread to other battleships and Hipper – Commander-in-Chief of the High Seas Fleet since the previous August – decided that it would be impossible to carry out the planned attack for the crews of the big ships were resolved not to go into battle because they believed that a battle now would cause the Allies to break off the Armistice negotiations.

In contrast, the crews of the destroyers and U-boats were ready to engage the enemy until the Armistice was signed. Throughout, the spirit of the little ships remained exemplary.

By a great stroke of foolishness, when the sortie of the fleet was cancelled, the Third Squadron of battleships was sent to Kiel which was already on the verge of revolution. For this feeling there was good cause: left-wing propaganda among the dockyard workers; defeatism brought into the base by the seamen and marines evacuated from the Belgian coast with, in the background, the general depression caused by the war news; long casualty lists and very short rations, all had their effect. Riots in Kiel soon spread all over Germany. On 8 November a delegation was sent to the headquarters of Marshal Foch to sign the Armistice and the Kaiser abdicated, on the next day fleeing to Holland, where his first demand was for 'a good cup of English tea'.

No neutral power had been willing to agree to the Allied request that it should take over the duty of guarding the German ships, and so they had remained throughout the winter and spring in the desolate Orkney harbour of Scapa Flow, whence they had been escorted by the British Grand Fleet while their fate was being decided. Each ship carried a German nucleus crew, no British personnel being permitted on board after the inspection which had been carried out following their arrival in British waters.

The interned fleet spent a miserable winter trying to eradicate traces of the mutinies which had taken place in October and November, while its commanding officer, Rear Admiral Reuter, was secretly planning the scuttling of the fleet. These two sets of circumstances were closely linked, for badly disposed seamen would almost certainly have tipped off the British had they had any inkling of what was toward. In the meantime the worst behaved members of the crew of *Friedrich der Grosse*, Reuter's flagship, passed their spare time roller skating along the steel deck over the officers' quarters aft and generally behaving badly, so that Reuter was obliged to shift his flag into the light cruiser *Emden*, where it remained until the end. His officers approached him with plans for scuttling the fleet, but he was obliged to reject them lest his plans might be leaked before he was ready. In fact, as early as January 1919, Reuter had discussed the scuttling with Commander Oldekop, his Chief of Staff, and he had begun to make his plans.

Neither Reuter's own government nor the British Admiralty bothered to keep him informed about what was going on at the peace conference and he had to make up his own mind as to what he was to do. He came to the conclusion that he would sink the ships only if the British attempted to seize them without the consent of the German government. 'Should our government agree to the peace conditions under which the ships were to be surrendered,' he wrote, 'the ships will then be handed over to the lasting shame of those who put us in this position.'

As the winter of 1918–19 wore on, so did the discussion at the peace conference as to the future of the German ships. It was easy for the Allies to decide that the ships should not go back to Germany, and during the discussions it was apparent that there was little real desire in Germany to retain them. Before the war the existence of the German navy had been very much a political matter, but now that there was a government of the Left no one in authority wanted to keep them, so that when Part V – the disarmament clauses of the peace treaty – was concluded Germany was permitted to retain only 8 obsolete battleships, 8 obsolete light cruisers and 32 destroyers and torpedo-boats, also obsolete. There were to be no submarines and no aircraft, naval, military or civil. The size of the vessels to replace obsolete ships was to be strictly limited, thus, armoured ships were not to exceed 10,000 tons, cruisers 6,000 tons and destroyers 800 tons.

Surrendered submarines were all broken up, except for ten which were incorporated in the French navy, while other powers were allowed to retain a number of boats for experimental purposes over a period of months. The largest types served as the basis for the design of the long range Japanese submarines of World War II.

Meanwhile, the Allies debated the fate of the ex-German surface ships. There was considerable bickering. An American spokesman led off by insisting that the ships should not go to Britain, for that would be a clear proof that Britain was preparing to attack the United States. He was wrong for at least two reasons: first, despite the clamour raised by the 'Big Navy' movement in the United States Britain had neither the intention nor the capability of fighting that country; in the second place, while the British did not want the ships themselves, they were anxious that no one else should have them either. 'No one else' meant, in this case, the French who had claimed the right to take over nine capital ships because their naval construction programme had had to be scrapped in 1914 since what was left of the French industrial effort, after general mobilisation, had been almost entirely devoted to the support of the army.

On 7 May the Allies finally presented the draft peace treaty to the Germans. A few minor changes were made at the last minute but on 20 June it was presented again to the Germans with an ultimatum due to expire the next day. The German Chancellor, the Socialist Scheidemann, refused to accept it and resigned. Gustav Bauer was appointed to succeed him and the ultimatum demanding the signing of the treaty was prolonged by two days. Reuter was not informed of this and continued to believe that 21 June remained the date fixed. He also held the view that in the last instance the German government would refuse to agree to the surrender of the fleet. In the middle of this confusion another hare had been started by an article in *The Times* of 16 June, stating that the German government was prepared to cede the interned ships – as well as those older ones which had been left behind in German waters – after they had been disarmed, in return for a financial settlement, presumably to be credited against the reparations account. On the basis of this information, Reuter sent a signal of protest to Berlin, demanding that he and his officers be recalled to Germany. The signal never reached its destination, probably because it was overtaken by the events of 21 June.

On the morning of that day the British First Battle Squadron of the Atlantic Fleet, commanded by Vice-Admiral Sir Sydney Fremantle, which had been stationed at Scapa as guardships, had gone to sea for exercises. Fremantle had been concerned that the Squadron was getting stale at Scapa where bad weather had prevented any training exercises, and with the first sign of good weather, which came on the 21st, he had decided to take his ships to sea. He had previously asked the Admiralty to give him a daily political appreciation of the situation so far as the signing of the peace treaty was concerned, but this was not done. Fremantle thought he had time to carry out some of the required exercises on 21 June and return to the Flow in time for the new hour and date of the expiry of the ultimatum, which was to be at 7 pm on 23 June.

All preparations had been made in the German ships after any unreliable hands had been sent back to Germany – men who might have informed the British of what was being prepared. Reuter stressed in his orders the fact that the crews could still not be trusted and that it was the officers who would have to be relied upon to watch for orders and carry them out. Kingston valves, fitted to discharge water from inside the ships, were reversed so that water would flood into them, submerged torpedo tubes were opened as well as condenser gratings, and all scuttles, doors etc. Inside the ships these apertures were to be jammed open, and there would be no lights inboard once scuttling had started – this being a measure to hinder British boarding parties to find their way about. Finally, Reuter distributed these orders through the fleet by means of the British drifter which carried communications around the Flow.

Fremantle and his squadron sailed out into the Pentland Firth at 9 am. An hour later Reuter signalled his ships to be prepared to scuttle and at 11.20 he gave the order. At noon the ship's bell of the *Friedrich der Grosse* began to ring loudly with single strokes. There was a certain amount of activity to be seen on the decks of the big ships. A junior officer in one of the German destroyers remarked later that this was the first time since they had arrived at Scapa Flow that anyone had been seen running aboard one of the big ships. At 2.20 a radio signal from a destroyer in the Flow reached Fremantle in *Revenge*:

'German battleship sinking.' This was followed a few minutes later by another:

'German ships sinking, some already sunk.'

Fremantle gave the order to return to the Flow, while the two destroyers left behind there and the trawlers did their best to cope with the situation, but there was very little that they could do. Even if they boarded the ships they were unable to find their way about in the complete darkness. In fact the only thing to do was to tow the ships that could be saved into shallow water, and this was done as far as possible, but capstans had been jammed to prevent anchors from being raised.

Preparations for sinking had been made in such a way that the ships were flooded so that almost all of them capsized. Only the *Hindenburg*, among the big ships, sank so that she rested upright on the bottom, while the *Baden* was taken in tow before she could sink and was beached in shallow water. *Emden* also sank in shallow water. Her crew had been observed by a guard trawler getting out her boats and loading them with the crew's belongings. Guessing what this meant, the trawler's skipper sailed slowly past the sinking light cruiser collecting the boats with boat hooks and towing them away. The *Emden*'s crew thereupon postponed their plans for scuttling the ship thus giving time for her to be towed aground.

The extraordinary events of this day were photographed with a thoroughness rare at that time. The captain of the depot ship *Victorious*, a disarmed battleship, recalled that he had on board a photographic rating, and he sent him off at once in a drifter. He returned that evening with a fine collection of pictures, the originals of which are preserved in the Museum at Kirkwall, but there was one picture which, unfortunately, the photographer did not get. On the forecastle of the *Baden* apparently alone in the deserted great ship, a solitary seaman, dressed in whites, was seen dancing a hornpipe.

10

The Coastal Motor Boats at Kronstadt, 1919

On 21 November 1918 the British light cruiser *Cardiff*, flying the flag of Rear-Admiral Edwyn Alexander-Sinclair, had been the centre-piece of the greatest spectacle of naval power the world had ever seen, when she led the major part of the German High Seas Fleet from the Firth of Forth into internment.

This moment, when the eyes of the world were upon her, was brief. On the next day, in company with the other ships of the Sixth Light Cruiser Squadron, *Cassandra*, *Caradoc*, *Ceres* and *Calypso*, together with nine destroyers and seven fleet minesweepers, she sailed for the Baltic – the first British warships, except for submarines, to enter that sea since 1914.

Cardiff's new task, and that of the ships under her command, was in stark contrast with the simple star role she had played on 21 November, for she was now running into a situation as dull, complicated and dangerous as any which had confronted the navy in the war which was just ending. To begin with the situation with which Sinclair's ships had now to cope was the emergency caused by the collapse of the three great Empires of Central and Eastern Europe, German, Russian and Austrian, and amidst their ruins there was civil strife ranging in intensity from rioting to full scale civil and international war. On the shores of the Baltic itself, a four-cornered struggle was going on in front of Petrograd, the city which had been known as St Petersburg until September 1914, had been capital of Russia until January 1918, and which was to become Leningrad in 1924.

In the winter of 1918–19 it seemed likely that the city would fall to the White, anti-Bolshevik forces of General Yudenitch, who was also on the verge of hostilities with the Estonian Nationalists, who had declared themselves independent of Russia, either Red or White, and were equally resolved to defend themselves against the Germans who, having lost the war and its spoils on the Western Front, were endeavouring to recoup their lossses in the Baltic states.

1. During the summer of 1893–4 a kind of slow motion civil war broke out in Brazil, with ships of the navy in the harbour of Rio de Janeiro opposing the government ashore. The two sides spent weeks in a state of stalemate, but eventually the principal ship on the side of the navy, the battleship *Aquidaban*, was torpedoed and sunk by a torpedo gunboat. This picture shows the *Aquidaban* after she had been raised and drydocked.

2. One of the earliest photographs of action damage sustained by a warship. *Kaiser*, the second flagship of the Austrian fleet at the battle of Lissa (Vis) 20 July 1866, flying the broad pennant of the second-in-command of the Austro-Hungarian fleet at Lissa, Commodore von Petz. *Kaiser*, a 90-gun wooden line-of-battle ship, hardly differs from Nelson's *Victory* save for primitive 800 hp engines. She followed the Austrian admiral Tegetthoff and his flagship *Erzherzog Ferdinand Max* through the Italian line. Tegetthoff rammed and sank the Italian *Re d'Italia*, while the *Kaiser* led the second Austrian division and, although a wooden ship, tried the same tactics against the Italian *Re di Portogallo*, but succeeded only in losing her bowsprit and figurehead as well as her foremast, main and mizzen topmasts and was set on fire.

3. In the early weeks of 1898 there was great tension between the United States and Spain, based on the American support of the Cuban guerillas in rebellion against Spain. The USS *Maine* was sent to Havana to protect American citizens and on the night of 15 February was sunk, following an internal explosion. American opinion blamed Spanish or Cuban extremists and the event brought nearer the United States declaration of war which took place on 25 April. Experience in the two World Wars of the behaviour of high explosives made an accident the most likely explanation of the disaster. This picture of the *Maine* was taken in the afternoon of the day she was destroyed.

4. Russian battleships sunk at Port Arthur. After a campaign lasting eleven months the Japanese army finally stormed positions in the hills around the great Russian naval base. This enabled them to bombard the Russian fleet in the harbour so that, one after another, four Russian battleships were destroyed. L. to R. *Peresviet, Poltava, Retvisan.*

5. *Audacious*. At the beginning of World War I no one was sure of the extent to which battleships were threatened by submarine or mine. Accordingly the British Grand Fleet spent most of the first three months of the war in Irish waters or those to the west of Scotland, but the danger to the Allies on the Western Front presented by the German attack at Ypres caused the Grand Fleet to be brought round into the North Sea. On the way the brand new battleship *Audacious* struck a mine laid by the German armed merchant cruiser *Berlin* and sank.

6. *Bouvet*. When stalemate had been reached during the winter of 1914–15 on the Western Front the Allies mounted a great attack on the Dardanelles to bring help to the Russians. Eighteen British and French battleships were concentrated for this attack. During the three months of February to May 1915 six Allied battleships were sunk by mine or torpedo. The first of these lost was the French *Bouvet*.

7. After the failure of the attack on the Dardanelles the North Sea again became the most important theatre of the naval war. On 31 May 1916 the German High Seas Fleet set a trap for the British Grand Fleet which failed but, nevertheless, British losses were heavier than the German. Three British battle-cruisers were sunk, the remains of one, *Invincible*, are seen here after an explosion had blown her into two parts which for several hours rested upright on the sea bottom.

8. Throughout the war the Allied and Austrian fleets watched each other
across the Adriatic. On the last day of the war two Italian frogmen entered
the Austrian naval base at Pola (now Pula) and sank the Austrian flagship,
Viribus Unitis.

9. *Majestic*. Another one of the six Allied battleships sunk in the Dardanelles. *Majestic* fell a victim to the German submarine *U 21* which, to the surprise of Allies' and Germans alike, had made the long journey from her home waters to the Dardanelles.

10. Under the terms of the Armistice which ended fighting in World War I, all U-boats were surrendered to the Allies while the sixteen biggest and newest of the German battleships and battle-cruisers were interned in British waters. Owing to a misunderstanding by the German admiral left with his fleet, together with much reduced crews, the order was given to scuttle all German ships, on 21 June 1919. The picture shows one of the two biggest battleships, *Bayern*, settling by the stern.

11. *Bismarck*, opening fire on HMS Hood, in the first moments of the long drawn out battle which ended with both ships being sunk (24–27 May 1941).

12. *Cavour*, one of the three Italian battleships sunk by aircraft of the British carrier *Illustrious* at Taranto, on 11 November 1940. This ship was sunk on two more occasions during World War II, first on 8 September 1943 when she was scuttled to prevent her falling into German hands at Trieste, and, having been salved a second time was sunk again by US bombers on 15 February 1945. At the end of the war she was raised again and, this time, scrapped.

13. Pearl Harbor. *West Virginia* sunk and *Tennessee* badly damaged by the Japanese attack of 7 December 1941. *West Virginia* as the outboard ship received the full force of the Japanese attack, from which she succeeded in protecting, to a great extent, the *Tennessee*.

14. *Yamato*. The Japanese battleship *Yamato*, then the biggest warship in the world, blows up and sinks after being hit by 10 torpedoes and 23 bombs from 386 US carrier-borne aircraft in an unsuccessful attempt to break up the American attack on the island of Okinawa on 7 April 1945.

15. The brand new flagship of the Italian fleet, *Roma*, was attacked and sunk by the Luftwaffe on 9 September 1943 by glide bombers. Two explosions split the ship in two, and she sank with a loss of 1,254 lives, including her admiral and her captain.

16. After the American victory at Leyte Gulf, the Japanese battle fleet ceased to exist as an organised force, being without oil fuel and air cover and the ships were laid up in Japanese ports. The *Haruna*, a Kongo class battleship of the Imperial Japanese Navy, on the bottom at Edashima Naval Base following an attack by US carrier-based aircraft in April 1945.

There were also hostilities spasmodically in progress between Poland and Germany, Poland and Russia (which had a civil war of its own), between Poland and Czechoslovakia and between Hungary and Rumania. With these, however, the British navy had nothing to do, except that, into the conflict between Germany and Poland, HMS *Concord* injected the personality of the world-famous pianist Ignace Paderewski and that of Lieutenant-Commander Bernard Rawlings, RN. Paderewski was to be chosen as first Prime Minister of the newly established Polish Republic, after long and acrimonious discussion between groups of Polish exiles. Lieutenant-Commander Rawlings found himself arranging for the evacuation by German forces of territory which had become part of Poland, and in this capacity negotiated with General von Blomberg, later Hitler's Minister for War, to the considerable surprise of both men.

The *Concord*'s passage to Danzig had been something remarkable, even in the history of the British navy, with the world's greatest pianist playing on a piano whose quality had been restricted by the amount to which the wardroom funds could run and the carefree handling which it had doubtless received during wardroom sing-songs.

Meanwhile Sinclair and his ships were on their way to Reval (now Tallin) in answer to an SOS from the Estonian government as the Bolsheviks were threatening Reval.[13]

On the way, off the island of Ösel, the *Cassandra* struck a mine in an unmarked field on the night of 4/5 December and sank, her back broken, after resting for a short time with her bow on the sea bottom and her stern standing out of the water.[14]

Mines, however, were not the only worry which faced the British ships, for almost nothing was known about the condition of the Russian Baltic Fleet. It had been reported in a state of almost continuous anarchy ever since the spring of 1917, but the British troops which landed in North Russia had met with several disagreeable surprises at the hands of the Red Army and Sinclair did not wish to take unnecessary risks by confronting his 6-inch gun light cruisers with Russian battleships carrying 12-inch guns. Two of these, *Petropavlovsk* and *Andrei Pervosvanni*, were based on Kronstadt, in seagoing condition, and supported by cruisers, destroyers and submarines.

The belligerent status of the Russian ships was by no means clear. Britain was not at war with Russia and did not wish to be, partly because of financial exhaustion and partly because there was an important amount of pro-Russian sentiment, mostly based upon a confusion of Bolshevism with left-wing Socialism and Liberalism. As a result, the members of the British government, with the

exception of Churchill, then Secretary of State for War, were determined not to send any British troops to fight ashore. They wanted
to keep losses to a minimum and feared that the troops might be
contaminated by Communist propaganda.

The French had had these worries on an even greater scale, for
the French fleet in the Black Sea had already been withdrawn
following a series of mutinies. As a result, the French senior naval
officer in the Baltic, Contre-Amiral Brisson, was unable to use his
ships to fight the Russians and similarly US destroyers, which had
escorted merchant ships engaged in famine relief, were told by
Washington that the merchant ships would have to rely upon the
British navy for protection.

At this time the Admiralty had obtained, with some difficulty, a
statement from the Foreign Office that 'Bolshevik men-of-war
operating off the coast of the Baltic provinces must be assumed to
be doing so with hostile intent and should be treated accordingly'.

This directive was put into effect for the first time on Boxing
Day 1918. Bolshevik troops were closing in on Reval, while Captain
B. S. Thesiger, with *Calypso*, *Caradoc* and five destroyers, was
harassing their advance by shelling the coast road along which they
were moving. The Russians had reacted to this by getting together a
collection of reasonably seaworthy warships, consisting of the
battleship *Andrei Pervosvanni*, the cruiser *Oleg* and the destroyers
Avtroil, *Azard* and *Spartak*, all under the command of F. F. Raskolnikov, formerly a sub-lieutenant of the Tsarist naval reserve, who
had been named 'Member of the Revolutionary War Council of the
Baltic Fleet'.

At 7 am on the morning of 26 December Raskolnikov signalled:
'I am going to bombard Reval.'

Ashore preparations were being made for a banquet to be given
in honour of the British crews, but it had to be postponed because
shells started to fall in the harbour and in the town. Out at sea a
warship could be seen steaming fast, her gun flashes twinkling. The
destroyer *Wakeful*, followed by the light cruisers, scrambled out
of Reval in search of the *Oleg*. *Calypso* failed to find her but met
the *Spartak* which she pursued in company with the *Wakeful*,
Vortigern and *Vendetta*. The Russian tried to get away but was
unable to shake off her pursuers as she could not exceed twenty-five
knots, while the British ships, according to the Soviet account of the
affair, were making ten knots more. This same Soviet account
commented blisteringly on the deficiencies of *Spartak*'s engine room
company who were unable to operate more than one of her two
turbines at a time. In addition the *Spartak* had been firing her forecastle 4-inch gun on so extreme an after bearing that her chart-

house was wrecked by the blast, her bridge was damaged, her coxswain was concussed and the charts which he was using were torn up or blown over the side, so that it was impossible to establish the ship's position.

The time was now 1.30 pm and about ten minutes later the muddy colour of the Russian's wake showed that she was in dangerously shallow waters. Almost immediately she ran aground, tearing off her rudder and both screws. She then lowered her flag, a British prize crew was put on board and she was towed back to Reval. As far as could be discovered this was the first occasion upon which an enemy surface warship had been captured by a British ship since the end of the Napoleonic wars.

Having enjoyed the postponed banquet that night, the next day Thesiger's men had a similar success when they met the *Avtroil*. Getting between her and Kronstadt they shot away her foremast and she too surrendered. Both these ex-Soviet destroyers were presented to the Estonian navy which, up to that time, had comprised only the gunboat *Lembit* of 875 tons, and a collection of small fishing vessels fitted as minesweepers.

One of the Russian prisoners taken was Raskolnikov, found hiding under some sacks of potatoes. He was afterwards exchanged for eighteen British officer prisoners and was entrusted by Trotsky with the suppression of the mutiny at Kronstadt in March 1921. With this success behind him he went on to the re-establishment of discipline in the whole navy, reintroducing into the service picked officers and petty officers from the Tsarist navy, under the supervision of reliable commissars. During the great purges of the last part of the 1930s Raskolnikov committed suicide.

With the capture of the *Spartak,* now renamed *Wambola,* and *Avtroil,* now *Lennuk,* the Estonians had the nucleus of a navy and to complete the picture they needed an admiral, whom they soon found in the person of Juhan Pitka, the head of a local salvage company, who took on his new role with enthusiasm and dash, literally as well as metaphorically, for he constantly led his ships about the Gulf of Finland at full speed – a very wasteful procedure as far as oil fuel was concerned. Pitka played a large part in giving his country the twenty years of freedom which it enjoyed between 1919 and 1939, and in addition to being presented by the British with the two destroyers was created KCMG.

The end of the war and the General Election in December 1918 which resulted in the return of Lloyd George's coalition were naturally the occasion for re-examination in London of relations with Russia and the Baltic states. On the one hand there was a call from the Left in Britain and France for a complete withdrawal from Russia and the neighbouring states, and on the other there was the belief that nothing could be done without an important military presence. Sir Rosslyn Wemyss, the First Sea Lord, supported this view saying that the only alternative to a British withdrawal at this time would be 'a land expedition of considerable strength'.

Instead of either of these alternatives a typical British compromise was chosen: the dispatch of two light cruisers and five destroyers under Rear-Admiral Walter Cowan to replace Sinclair and his ships, which were judged to have done their stint.

Cowan, one of the most notable and successful in the history of Royal Navy eccentrics, may be said to have had his life bracketed by two awards of the Distinguished Service Order, one as lieutenant at the age of twenty-seven after taking part in Kitchener's Nile campaign, culminating in a dash to Fashoda to head off a French expedition designed to annexe what afterwards became the Anglo-Egyptian Sudan, while the second award came in 1944 at the age of seventy-two for gallantry, determination and undaunted devotion to duty as Liaison Officer with commandos in the Italian campaign and among the islands on the Dalmatian coast.

During his career he had taken part in campaigns in West Africa, East Africa and South Africa, all on land, the latter including two years without the authority of the Admiralty, which decreed that this time should not count for promotion. He spent World War I in the Grand Fleet, successively as captain of a battleship, a battle-cruiser and, finally, as Rear-Admiral in command of a Light Cruiser Squadron.

Cowan was only five feet six inches tall and possessed a very quick temper, so that it was of great importance that he should always have a flag captain who could calm him down. The Second Sea Lord's office did not always make the necessary arrangements for this, with the result that Cowan's reputation as an officer whose ships were sometimes plagued by mutiny had an element of truth. On the other hand, when he had flag captains of the ability of Andrew Cunningham and Charles Little an excellent 'doubles pair' was the result, while Cowan excused himself for his temper by saying that when he had had a squadron he made the mistake of expecting too high a standard of discipline.

Another defect stressed by Captain Bennett in his book *Cowan's War* (1964) was that 'Cowan's early association with the Army had imbued him with the too long a-dying tradition, exemplified by Lucan's failure to remove Cardigan from his command after Bala-clava, that incompetence should be overlooked "in an officer of good family who rode a good horse".'

It was in World War II, when he was over seventy, that Cowan's career became most spectacular. He made desperate and ingenious efforts to get up with the fighting in the Western Desert, and finally attached himself to the 18th King Edward's Own Cavalry in the Indian Armoured Corps.

One of those to whom Cowan turned for help in getting him to the war was Sir Roger Keyes, who had led the raid on Zeebrugge in April 1918 and, in the first part of World War II, was head of Combined Operations. Keyes was sympathetic to Cowan's importunities, saying: 'Walter only wants to die for his country', which Walter very nearly did, but instead was taken prisoner at Bir Hakim, released on account of his age, volunteered and was accepted for commando operations in the islands of the Dalmatian coast, surviving the war and dying peacefully at the age of eighty-five in 1956.

The winter of 1918–19 and the first months of spring 1919 passed with low level political and military developments, as none of the Powers involved in the Baltic had sufficient forces to make an all-out bid to control the situation, but on 25 April came news of a fresh Russian move. The ice at Kronstadt had melted and the Red Fleet was at sea, with an estimated strength of the two battleships, *Petropavlovsk* and *Andrei Pervosvanni*, two or three cruisers, twelve destroyers and perhaps as many as seven submarines with minor patrol craft and minesweepers. Confronting this force was Cowan with two light cruisers and six destroyers. It was decided to send reinforcements from the United Kingdom, but there was to be no question of these including battleships to match the two Russians, for British battleships were not to be risked. Instead, in addition to more cruisers and destroyers, submarines, coastal motor boats and minesweepers were dispatched. Of these it was the CMBs that were to settle finally the outcome of the British navy's Baltic campaign. Cowan, disappointed in his desire for battleships, considered the possible use of aircraft to disable the Russian heavy ships, the range and accuracy of whose 12-inch guns very soon impressed the British as they had impressed the Germans at the battle of Moon Sound in October 1917. In that action the recently completed German battleships of the *König* and *Kaiser* classes had found themselves outranged and outshot by a pair of pre-dreadnoughts (*Slava* and *Grajdanin* formerly *Tsarevitch*) dating from the turn of the century. On 30 May, following a brush with the *Petropavlovsk*, 'Cowan', says Captain Bennett, 'was sufficiently impressed bȳ her "heavy and well-disciplined fire" to argue the need to take the Bolsheviks' naval efficiency more seriously than hitherto.' One immediate measure was to establish a forward base at Biorko Sound, a desolate anchor-

age thirty-five miles from Kronstadt. It was painfully reminiscent to
the British of Scapa Flow, but ships based there were assured of
being able to react quickly to any Russian sortie.

Cowan, in a dispatch to the Admiralty, pointed out that his much
smaller force would be constantly at the mercy of the Russians,
unless the latter could be hemmed into Kronstadt by mines, or their
principal ships damaged by bombs, torpedoes or gunfire.

The Admiralty replied, however, that the present policy of His
Majesty's Government remained in force. This precluded offensive
action against Kronstadt by monitors, coastal motor boats or bombing aeroplanes. Torpedo-carrying aeroplanes could not be used at
Kronstadt as the torpedoes could not be dropped in less than ten
fathoms (a point which was to be of vital importance in considering
the planning of the raids on Taranto in 1940 and on Pearl Harbor
in 1941).

The only immediate help that the Admiralty could send was the
minelayer *Princess Margaret*, a former Canadian Pacific passenger
steamer, with 300 mines on board, and an intensive programme of
minelaying was begun.

At this time the British introduced a new weapon into the struggle
for command of the Gulf of Finland, known variously in the years
to come as CMBs (Coastal Motor Boats), MTBs (Motor Torpedo-
Boats) and FPBs (Fast Patrol Boats) in the British navy, PT boats in
the US navy, S-boote (Schnell-boote, fast boats) in the German navy
(E-boats to their enemies) and MAS (Motorscafi Anti-sommergibile,
anti-submarine motor boats) to the Italians who had early taken the
lead in their development.

Cowan soon saw that these boats could be used offensively in the
latest development in the waters around Kronstadt. On the southern
shore of the Gulf, opposite Kotlin Island upon which Kronstadt
stands, was Fort Krasnaya Gorka, a Bolshevik strong point whose
troops mutinied and began shelling Kronstadt. To counter this and
to prepare for the recapture of Krasnaya Gorka, the two Russian
battleships emerged from Kronstadt and began a counter-bombard-
ment of the mutineers. Cowan saw that the enemy were thus ex-
posing themselves to attack by the CMBs, believing themselves
protected behind their minefields, and not realising that the very
shallow draft of the CMBs enabled them to cross the minefields
without danger. The peculiar circumstances of the quasi-war which
was in progress made it necessary for Cowan to obtain permission
from London to carry out an attack on the battleships. London
gave an answer which freed the British government from responsibility and left it to Cowan: 'Boats to be used for Intelligence
purposes only unless specially directed by Flag Officer.'

Cowan, being the Flag Officer concerned, then said to Agar, who had brought out from England two of the smaller CMBs – the forty-footers – that he couldn't direct him to attack 'but', added Agar, 'if I did I could count upon his support. I left at once!'

On that night, 16 June, Agar and his two boats, CMB *4* and *7*, set out to attack the two battleships, but *7* hit a submerged obstruction, broke her propeller shaft and had to be towed home by *4*. Next morning the battleships had gone back to Kronstadt, having been replaced by the cruiser *Oleg*, which was continuing to shell the rebel fort.

That night Agar decided to try for the *Oleg* by himself in *4*, accompanied by a crew of two, Sub-Lieutenant Hamsheir, RNR, and Chief Motor Mechanic M. Beeley, RNVR.

The CMBs were too small to mount torpedo tubes, so that the method of firing the torpedoes, which were carried in a trough on deck aft, was to point the boat at the target and expel the torpedo backwards by a weak explosive charge which also started its motor going forward. Meanwhile the CMB itself, moving at full speed, swerved out of the way of the torpedo.

On this occasion things did not go quite so smoothly. The torpedo's discharge cartridge was prematurely exploded and another had to be fitted within sight of the Russians who, however, failed to notice the CMB in the dark. Fitting the new cartridge seemed to take hours but eventually it was done. Agar worked up to full speed, tore through the enemy destroyer screen and fired his torpedo. A minute later there was a thick column of smoke from the *Oleg* which had now opened fire, together with the destroyers and the forts ashore. The CMBs had 350–375 hp aero engines, which made so much noise that the sound of the enemy guns could not be heard, but the gunflashes were only too visible and the shell splashes drenched the crew of Agar's boat as she drew out of range. Next morning the *Oleg* could be seen lying on her beam ends with her starboard side awash.

Encouraged by this sinking, Cowan began to plan a CMB attack on the big Russian ships lying in Kronstadt harbour itself. The most important targets there were the two battleships, *Petropavlovsk* and *Andrei Pervosvanni*. In addition there was the submarine depot ship *Pamiat Azova*, an obsolete armoured cruiser whose stores and workshops were vital to the maintenance of the Russian submarines, and the armoured cruiser *Rurik*, regarded as a prime target because she was believed to have 300 mines on board, the explosion of which would wreck most of the dockyard. There was also a second-class cruiser, either *Diana* or *Avrora*, and a destroyer, acting as guardship off the harbour, might be an extra dividend.

In addition to its forts and mines the harbour was protected by a heavy breakwater, with a single gap in it some fifty yards wide. Across this gap a boom had been placed, which would have to be broken before the seven CMBs could enter the middle harbour and get to work. Between them the boats carried twelve torpedoes; CMB *4, 24, 72* and *79* having one each and *31, 62, 86* and *88* two. Each boat had a complement of three to five men, reinforced on this occasion by Finnish smugglers acting as pilots through waters they knew well. Another expert, who was getting to know these waters well, was Agar who accompanied the expedition in CMB *4*. The boats' course lay along the north shore of Kotlin Island, then through the gaps in the chain of forts which stand in the sea barring the way up the Neva, followed by a slow U-turn so that the boats came up to Kronstadt from the east and south. Overhead the RAF was making as much noise as they could to drown the sound of the boats' engines, but there was nothing to conceal the sheets of flame which flared out of the boats' exhausts. However, the defences of Kronstadt were exclusively occupied with the RAF. Searchlight beams swung to and fro while twenty-four bombs came whistling down, an alarming sound but the heaviest weighed only 112 pounds and were not nearly powerful enough to damage the massive fortifications of Kronstadt and the outlying islands. However, the raid made the garrison keep their heads down, and although the British planes soon ran out of bombs, they continued the good work of distracting the attention of the Russians by flying down the searchlight beams, machine-gunning as they came.

The CMBs were now approaching the entrance to the harbour. They formed up in two groups. First came *79, 31* and *88*, with Commander C. C. Dobson at the head of the three boats aboard *31*. The boats then slowed down to reduce noise and damp down their bow waves. The Russian guardship, the destroyer *Gavryil*, gave no sign of noticing that anything was going on. Lieutenant W. H. Bremner, in *79*, was to deal with the boom at the harbour mouth, but on arrival there found that the boom was not in place on that night.

Taking advantage of this Bremner opened up to full speed, then swerved to starboard so that the *Pamiat Azova* lay squarely across his bows. He fired his torpedo which hit and the enemy ship rolled over on her side. The next boat was Dobson's, who had as his target *Petropavlovsk*. He swung to port and, as Agar wrote afterwards: 'His attack was a more difficult manoeuvre; he had to stop one engine, turn the boat and gather speed quickly again before firing. This requires great judgment and coolness, but he did it; his torpedoes found their mark.'

The explosion of these three torpedoes alerted the Russians to what was going on and every gun in the harbour that could bear now opened fire. The third boat in, Dayrell-Reed's *88*, was hit and her captain mortally wounded. The boat sloughed off course; Steele, the First Lieutenant, steadied her course for the other battleship, *Andrei Pervosvanni*, which was lying alongside the already stricken *Petropavlovsk*. Steele fired both his torpedoes but before they exploded there were two loud crashes as the *Petropavlovsk* was hit.

'Then there was another nearby,' said Steele afterwards. 'We received a great shock and a douche of water. Looking over my shoulder I realised that the cause of it was one of our torpedoes exploding on the side of the other battleship. We were so close to her that a shower of picric powder from the warhead of our torpedo was thrown over the stern of the boat, staining us a yellow colour which we had some difficulty in removing afterwards. Missing a lighter by a few feet, we followed Dobson out of the basin. I had just time to take another look back and see the result of our second torpedo. A high column of flame from the *Andrei Pervosvanni* lit up the whole basin. We passed the guardship at anchor again. Morley (*88*'s mechanic) gave her a burst of machine-gun fire as a parting present and afterwards turned to see what he could do for Reed.'

The guardship *Gavryil* was still there, untouched because *24*'s torpedo had missed, diving below its target. *Gavryil* reacted so promptly and so accurately that a shell hit *24* and sank her at once.

Brade's *62*, coming in, collided with Bremner's *79* on her way out after having sunk the *Pamiat Azova*. The two boats were locked together but Bremner was able to jettison the explosives he had on board, originally intended for blowing a hole in the boom, then clamber out and scuttle his crippled boat. Brade now tried to torpedo the *Gavryil* but failed as Napier had done, his torpedo also underrunning the target and *Gavryil* sank him at once, and then lowered a boat to pick up the survivors.

Of the two remaining CMBs, Howard's *86* broke down but was towed home by Bodley in *72*, who was unable to take part in the action, a shell splinter having wrecked the firing mechanism of his torpedo.

In all, three boats had been lost and 6 officers and 9 ratings killed, while 3 officers and 6 ratings were prisoners of war. These were eventually returned to England with no serious complaints about their treatment by the Russians.

Two VCs were awarded, to Dobson and Steele, four officers received the DSO and eight the DSC, while all the ratings received the CGM (Conspicuous Gallantry Medal). In Britain and in the fleet there was great enthusiasm over the attack on Kronstadt but, as Captain Bennett points out, this was not shared by the British government which, at that time, was secretly negotiating with the Russians for a peace agreement, a process made more difficult by the fact that there was not, officially, any war for the peace to bring to an end.

The *Petropavlovsk* was raised in time to take part in the Kronstadt rising of March 1921, although her part was strictly limited by the fact that she was frozen in the ice at Kronstadt with her sister-ship *Sevastopol*, and was thus unable to carry out a bombardment of Petrograd as the mutineers planned. She was rearmed, re-boilered and renamed *Marat* between 1926 and 1933. She took a modest part in the Russo-Finnish war of 1939–40 but when the German invasion came in the following year her 12-inch guns were of the greatest importance in the defence of Kronstadt. Hit by several bombs and her forward turret knocked out, the ship sank in shallow water for the second time in her career, but the three remaining turrets continued in action throughout the siege of Leningrad.

11

'What a Wonderful Feat of Arms!'

Churchill on the sinking of the *Royal Oak*, 1939

By March 1939 it had become clear that the end of the Czech crisis of the previous autumn had not put an end to the likelihood of war. In August the German navy began to prepare for immediate war at sea. On 5 August the fleet oiler *Altmark* sailed for Port Arthur, Texas, to load diesel fuel for the pocket battleship *Admiral Graf Spee*, which left Wilhelmshaven for the South Atlantic on 21 August and there awaited orders to begin her career of commerce raiding. Two days earlier, fourteen U-boats had left Germany for positions in the North Atlantic and two more boats sailed on 23 August, the day after another fleet oiler, *Westerwald*, had sailed for a rendezvous with the pocket battleship *Deutschland* (later re-named *Lützow*) off the southern tip of Greenland.

The first general warning of the immediate probability of war went out to German merchant shipping on 25 August, and two days later it was followed by another general signal ordering German merchant shipping to do everything possible to reach their home ports or the ports of a friendly or neutral state within four days.

The world held its breath and hostilities began at first light on 1 September. The German battleship *Schleswig-Holstein* was at that time on a courtesy visit to the Free City of Danzig and at 0445, without warning, opened fire on the Polish base of Westerplatte on the edge of the town, and then put ashore a landing party which failed in an attack on the Polish position.

Next day German light forces commanded by Rear-Admiral Lütjens, who was to lose his life in the *Bismarck*, shelled other Polish positions on the Gulf of Danzig, and the Polish torpedo-boat *Mazur* was sunk by dive-bombing, the first warship on either side to be sunk in World War II.

On 3 September the war spread from the Gulf of Danzig to the North Sea and the North Atlantic, when Britain and France declared war on Germany and, on that same afternoon, *U 30* (Lieutenant Lemp) torpedoed and sank the Anchor liner *Athenia* with 1,300 people on board, 110 of whom lost their lives.

This action of Lemp's, which was against orders, altered in a few minutes the whole picture of the war at sea. Attacks on passenger liners had been forbidden by Berlin because Hitler still believed that the Anglo-French declarations of war were a kind of bluff and that the two allies would agree to make peace as soon as it was clear that they could not save Poland. The German authorities at once denied Lemp's responsibility and there was a collection of half-wits and German propagandists who took up with vigour the slogan 'Churchill sank the *Athenia*'.

Apart from spreading this kind of confusion, Lemp's action had an immediate effect on the decision of the British Admiralty, taken during the years between the wars, that the convoy system would not be introduced for ordinary merchant ships on the outbreak of war, despite the essential part convoys had played in defeating the unrestricted submarine warfare campaign of 1917–18. For this decision the Admiralty had given a number of reasons, all of which had been advanced and disproved against the introduction of convoys in 1917, except for the shortage of anti-submarine craft.

This shortage of 1939 was exemplified on 17 September when the aircraft carrier *Courageous* was sunk by *U 29*, while escorted by only two destroyers. Two other destroyers had been detached to hunt U-boats reported elsewhere in the Western approaches.

American naval observers at the time commented that, whereas *Courageous* had an escort of four destroyers, it was the practice of the United States navy to provide an escort of eight, but the tasks of the British navy in feeding the country and supplying forces and allies overseas did not permit this apparently lavish use of resources.

Meanwhile, as early as 6 September, the first steps were taken by which the U-boat arm could secure its most sensational single success of either World War. On that day a photo-reconnaissance Heinkel 111 of the Luftwaffe flew over Scapa Flow taking pictures. Examined in Kiel they showed, to the great surprise of the Germans, that not all the entries into the Flow itself were blocked by nets or by sunken ships. It was clear that it might be possible for a daring and skilful submarine captain to take his boat right into the principal British naval base. This intrusion had been attempted twice during World War I, in its first months and in its last, and both attempts had failed.

The undefended state of Scapa Flow in 1939 arose from the decision that a move such as the installation of fixed defences, protection by radar chain etc. would be regarded by the Germans as a threat and would also cause alarm amongst the inhabitants ashore, so that nothing was done until war was at hand, and the work had not been completed six weeks after the beginning of hostilities.

This state of affairs almost completely reproduced that of 1914. Scapa was the obvious place upon which to base the blockade of Germany, but it was completely undefended and the fleet was unable to lie there with any degree of safety, so that in August 1914 Jellicoe kept the Grand Fleet in harbour for only one complete day and in the following month for only six complete days. Things were just as bad in October when, on the 27th, the need to keep the fleet on the move to counter a possible move by the German fleet to support the army in the battle for Ypres led to the loss of the brand new battleship *Audacious* by a combination of mine and internal explosion.

Things were much the same a quarter of a century later, and Scapa had to be abandoned as the main base of the Home Fleet until its defences were placed in order. Meanwhile the fleet moved between the Clyde, Rosyth and Loch Ewe in north-west Scotland.

By Sunday 1 October Dönitz and his staff had finished their examination of the photos taken by the Heinkel, and had decided that it would be feasible for a U-boat to enter the Flow and attack whatever it could find inside. Dönitz accordingly sent for one of his U-boat commanders, Günther Prien, and asked him if he thought that he would be able to do the job. He was to have forty-eight hours in which to make up his mind but, Dönitz told him, no one would think the worse of him if he decided that it could not be done.

Forty-eight hours later Prien was back again. After hours of working with charts and tide tables he had decided that it could be done, but only on the surface. There was no prospect of getting in and out submerged. A nautical almanac supplied the phases of the moon and Prien and his boat, *U 47*, were ready.

Prien himself was a character whose life story was a perfect reflection of the age and place in which he grew up, Germany between the wars. His was a one parent family; the Priens' only means of livelihood was provided by Frau Prien touring the villages around Leipzig buying lace made by the local cottagers and reselling it to the smart shops of Leipzig. These were the days of great German inflation, when German money had no value at all and foreign money could buy the moon. Young Prien, then fifteen, hired himself out as a courier and general odd job man to foreigners visiting the

Leipzig Fair, and managed to collect enough foreign money in tips and fees to finance his heart's desire, which was to go to sea after he had successfully graduated from a college for merchant service cadets. He did this and obtained a berth as Fourth Officer. But Prien was not free from the curse of the Depression which had been roaring through world shipping and which had left him in 1931, twenty-two years old, without a ship.

At this time the Nazi party, still in opposition, was running a series of establishments known as *Landjahrheime*, where unemployed youths were given rugged board and lodging and put to work in agriculture, road building and other simple unskilled jobs. Prien was appointed to run one of these camps. Just before Hitler came to power the German government began cautious preparations to violate the disarmament clauses of the Treaty of Versailles. The navy needed more men and in the summer of 1932 it was discreetly made known that junior officers of the merchant service were needed as warrant officers in the *Reichsmarine* (the title of the navy of the Weimar republic). Prien applied, was accepted and four years later was already First Lieutenant of a U-boat. When war began he had his own boat, *U 47*, and by October was on the threshold of Scapa Flow. He briefed his crew and promised them a harvest of aircraft carriers, battleships and cruisers; in short, a rich selection of the most important units of the British Home Fleet. Preparations were then made to scuttle *U 47* in case it became necessary, and the crew then lay down to rest. Even if they did not sleep, by lying down they would be using less oxygen than had they been moving about.

At 4 pm hands were called to dinner, for which the cook had made a special effort, smoked mutton, greens, a pudding and coffee. Then they surfaced. It was a moonless night, but there was a splendid display of the northern lights, continually changing, with a dry, crackling sound. To enter the Flow Prien headed into a narrow channel between the mainland of Orkney and the island of Lamb Holm, which had been partly sealed by blockships sunk in the channel. Prien's first attempt to pass through this gap failed, for he took a wrong turning. Correcting his course he continued on his way and, suddenly, the boat was caught in the headlights of a car passing along the shore road. To this day the identity of the car and its driver have never been discovered. Ahead lay two blockships, filled with concrete and scuttled. The gaps between them and the shore were closed by wire cables and chains. Onto one of these cables *U 47* ran with a rasping sound and the cable itself forced her down so that she struck the bottom and scraped along. Prien backed away, adjusted his trim and the boat rose over the wire. One more check and then Prien was able to pass the word down the open conning tower hatch:

'We're in!'

Slowly they entered the great harbour which measures some sixty square miles. But where they had expected to see the British heavy ships there was nothing at all. The Home Fleet had vanished. The northern lights caused Prien to comment, 'It is disgustingly light, the whole bay is lit up.'

Moving slowly northwards towards Scapa Bay Prien then saw what he thought were two capital ships, a battleship of the 'R' class and a battle-cruiser. It was now nearly an hour since he had entered the Flow and he decided to attack, preparing to fire his four bow torpedoes at the two ships which seemed to be asleep. One torpedo failed to leave its tube, the others set off on their three and a half-minute dash across the harbour. One hit the battleship and the leaping fountain of water could be clearly seen. The Germans held their breath but there was still no reaction from the British ships. Prien had a quick, urgent exchange of ideas with Endrass, his First Lieutenant. Should they try to leave the Flow at once, or should they go further out in the harbour themselves, reload their tubes and come back again for another attack, as Endrass urged. Prien agreed with him, fired his stern tube with no effect and then reloaded, at 0116 firing three more torpedoes.

On board the British battleship, which was the *Royal Oak*, the flagship of Rear-Admiral H. E. C. Blagrove, Second Battle Squadron, the first explosion had scarcely been noticed, many men sleeping through it.

Admiral Blagrove, Captain W. G. Benn, *Royal Oak*'s captain, and other officers hastily discussed what had happened while investigation was being made in the forward part of the ship where the explosion was believed to have occurred. The paramount danger to the fleet at that time was considered to be from aircraft, and the first explanation was that a bomb had been dropped. Air defence stations had been kept manned around the clock and they could report nothing. The next idea was the possibility that the explosion had occurred inside the ship, in the CO_2 machinery space, the centre of the ship's refrigeration. Then it was discovered that the ship was unmoored, the anchor cable having been snapped. Subsequent investigation pointed to its having been hit by a torpedo.

The proccupation with the danger of air attack had led to as many men as possible being kept below in the ship, which meant that they had great difficulty in escaping when, at 0116, two more of Prien's torpedoes hit the ship.

Several men, either on watch or still discussing the events of the night, saw a splash of spray, two towering water spouts and then felt and heard two heavy explosions amidships, near the boiler and engine rooms and the dynamo room on the starboard side. Within a few minutes the ship was listing heavily and quickly. Officers and men began, with great urgency, to throw overboard anything that would float. The order was given to signal by searchlight, asking for immediate assistance, and the answer came back that it was impossible, all power had gone and with it all the lights within the ship. A sheet of orange flame came up from below the deck and licked around the starboard side of the funnel casing, and then up the casing itself. The list on the ship grew greater and thick black smoke and a smell of burning cordite covered the afterpart of the ship, while the upper deck was drenched in fuel oil from ruptured tanks.

From below came the cries of suddenly alarmed men and the sounds of steel bending and bulkheads rending. Auxiliary lighting and power came on in different parts of the ship and, throughout, queues of men were forming up to climb the various ladders to the upper deck and safety, although many hatches were found to be closed for reasons of damage-control. The auxiliary lighting now proved intermittent. In one part of the ship, in complete darkness, a warrant officer was striking matches to show men the escape route. Some men were able to help others in their search for safety, others ran into closed doors or hatchways, turned back and then met, head on, another column of men trying to get along, and all the time the ship was listing further. Within a very few minutes she would turn right over and then there would be no way out at all.

Meanwhile other men, obedient to their orders as to what was to be done in case of an air raid, were still trying to get below, seeking the protection of armour.

Fierce fires, apparently started in the dynamo room on the starboard side amidships when one of Prien's torpedoes exploded, were followed by what seemed to be fireballs which chased along the alleyways of this part of the ship, consuming men as they met them, while the force of the explosion and the heat of the fire started rivets and through the gaps in the plating came more flames.

The ship was now turning over faster. Someone had a last glimpse of the Admiral refusing a lifebelt.

The *Royal Oak* went, with a series of roars and rumbles and crashes, as everything that could do so tore itself loose. The sea was covered with oil, in the midst of which were the heads of struggling men. A drifter called *Daisy* had a narrow escape from being caught up with the sinking ship, escaped and then went to work picking up men. As the battleship turned over she made a great splash and there was a sigh of escaping air. She lay on her starboard side, then turned further over. A searchlight suddenly came on and swept across the sea, showing the *Royal Oak* by now upside down with men walking about on her bottom.

Then she went.

Prien had believed, since this operation had been mooted, that, while it would be comparatively easy to enter Scapa Flow, it would be very difficult, if not impossible, to get out again. Now the time came when it had to be tried. Still on the surface he made his way back to Lamb Holm, a slightly different way through the obstacles this time, dodging bits of a block ship that had broken up and then, after two hours in Scapa Flow, home to a hero's welcome.

Behind him he left a badly shaken naval base. The complete surprise secured by Prien was reflected in the fact that life-saving efforts had been almost non-existent when the *Royal Oak* sank, save for the *Daisy* which happened to be on the spot and the *Pegasus* which was a mile away. This ship was taken by Prien to be another battleship and claimed by Goebbels to be *Repulse* and damaged. In fact she was not *Repulse* but a former merchant ship which had been completed in 1914 as the seaplane carrier *Ark Royal*, the first ship of her type in the world and the first of her name since the *Ark Royal* which had been the flagship of Lord Howard of Effingham during the campaign of the Spanish Armada.

The next morning after the sinking of the *Royal Oak* another blockship arrived at Scapa, to be sunk and block the gap through which *U 47* had entered and left the harbour.

As for the big ships which Prien had promised his men as targets – 'aircraft carriers, battleships and cruisers' – they had left Scapa on 8 October, in an attempt to intercept the first sortie of the war by a big German ship. The battle-cruiser *Gneisenau*, flying the flag of Admiral Boehm, and the light cruiser *Köln*, with an escort of nine destroyers, were steaming up the coast of Norway seeking any British light forces which they might destroy, after which they planned to return home, luring after them important British forces into concentrations of submarines or aircraft.

Admiral Forbes, C-in-C Home Fleet, did not succeed in catching the Germans; German bombers did try to attack the British Humber Force of two light cruisers and nine destroyers, also without success. He then decided to take his fleet, less *Royal Oak*, to Loch Ewe, which he considered safer than Scapa.

In all 833 men lost their lives in the *Royal Oak*, and 424 were saved. It is difficult to pick one's way through all the things that had gone wrong. The preoccupation with the danger of air rather than submarine attack led to mistakes about damage-control.

Obviously measures dealing with torpedoes were different from those necessary for bombs, but not only were men kept below when they would have had a much better chance of survival had they been on deck, but many damage-control measures within the ship, such as the more rigorous measures to check the closing of water-tight doors, would not have made the sinking of the ship so easy.

The *Royal Oak* was an old ship, completed in 1916 and never thought worth a really thorough reconstruction, such as that given to the preceding *Queen Elizabeth* class of battleships. If the war had not begun, it was intended to scrap the *Royal Oak* and her sister-ship in 1942. Her design suffered from the preoccupation of the Liberal government of 1906–15 with economy and arms limitation and the subsequent decision, at a time when all the principal navies were increasing the size of their battleships, that Britain should cut the size of those which she was laying down in 1913–14 from 27,500 tons to 25,000 tons, which meant a reduction in the amount of armour allotted to the protection of this class. The Germans replied to this by increasing the size of their new ships from 25,000 tons to 28,000 tons.

Prien and his men got back to Germany for the heroes' welcome they certainly deserved, despite Goebbels' claim that they had also torpedoed the battle-cruiser *Repulse*. Dönitz protested against the making of this entirely unfounded statement on the grounds that it would be bad for the morale of the navy, but he never succeeded in getting Goebbels to moderate his claims.

Prien was to go on to achieve further successes, all against merchant ships. Between the outbreak of war and 8 March 1941 when he himself was sunk, he sank twenty-eight ships, excluding *Royal Oak*, totalling 164,953 tons. This total made him one of the most successful of the German U-boat commanders, during what the German navy called 'The Happy Time' when British anti-sub-marine warfare was in its earliest stages of development, both as regards material and technique. It was during this period that the Germans sank 217 merchant ships, totalling 1,100,000 tons, losing only forty boats themselves, never having more than ten boats operational at a time, and sometimes only two. Before the war began Dönitz had calculated that 300 boats would be needed to assure victory at sea.

Prien was one of the 'Big Three' of the U-boat commanders in 'The Happy Time', the others being Kretschmer and Schepke. Be-tween them these three men sank nearly 600,000 tons. All three were sunk within a period of ten days, only Kretschmer surviving, as a prisoner of war. In his Canadian prisoner of war camp at

Bowmanville, Ontario, he organised a clandestine system of communications with Germany by which he was able to arrange for a U-boat to enter the Gulf of St Lawrence to pick up some German escaped prisoners of war. The U-boat arrived but in the meantime the prisoners had been recaptured. The triple loss of these men meant, as Captain S. W. Roskill wrote, the end of the waging of war by brilliant individual aces and that reliance, in the future, would have to be placed on team work between groups of boats.[15] The 'wolf pack' era was at hand.

It is interesting to discover that, although Prien was clearly one of the most successful of German submariners, his reputation as a commander was not as high as that of a number of his fellow captains, although his courage and skill as a ship handler were universally recognised. There was a certain amount of feeling that a man who could say, as Prien did, that he preferred a good convoy exercise to a spell of leave was too inhuman to make a really first-class leader of men.

12

The Coming of the Carrier

Illustrious, Taranto, 1940

When Clement Ader, the French engineer, designed the first aircraft carrier in 1894, the only aircraft in existence at that time which it could have carried, had it ever been built, were spherical ballons and, possibly, the 'electric dirigeable', *France*.

Ten years later a freighter carrying a spherical balloon for scouting purposes accompanied the Russian fleet to its doom at Tsushima, but there is no record of its ever having accomplished anything. By that time the Wright brothers had flown; by 1908 their demonstrations at Rheims had convinced a sufficiently influential collection of senior and medium rank officers in the principal armies of the world that heavier than air craft, as aeroplanes were sometimes styled in those days, had an important part to play in future land warfare. A little later men's mechanical abilities progressed a stage further so that aeroplanes would be able to intervene in naval warfare as well, but until 1911 the only development in that direction had been the German rigid airships of the Zeppelin type. In that year, however, a plane had flown off a temporary platform built on the deck of an American cruiser, and in the following year the same experiment was carried out from the deck of a British battleship. These tests had been made from ships lying motionless in harbour, but to be of real value in naval warfare it would be necessary for planes to be with the fleet at sea at all times. To achieve this took the British navy four years – the four years of World War I. The first stage occupied most of the two years 1914–16, at the beginning of which a number of ex-cross Channel steamers were equipped, in the first instance, with canvas hangars and carrying seaplanes – planes with floats instead of wheels. These were lowered onto the water and picked up at the end of their flight, a lengthy and dangerous process when carried out in waters which might contain enemy submarines. This danger was reduced and the process speeded up by the introduction of ships with decks from which planes could be flown, although they still had to land on the sea to be taken on board again.

The third and final phase of development came in October 1918, with the completion of the first real aircraft carrier. This vessel, laid down as an Italian liner, was completed as HMS *Argus*. She had no funnels or masts and a flush deck, with a small bridge and charthouse mounted on a lift so that it could be raised or lowered as required when planes were landing or taking off. The idea of a mastless, funnelless carrier may have seemed obvious, but it was not popular, and a single, big funnel on the starboard side of the ship became a practically standard fitting for carriers of all navies except the Japanese, apparently because pilots liked to have the funnel by which to measure their height above deck. In addition, if there was no funnel, gases from the boilers emerged from the stern and created dangerous turbulence.

By the time World War II began the British navy possessed one brand new carrier of 22,000 tons, the famous *Ark Royal*, a smaller one, *Hermes*, of 10,000 tons, three carriers that were converted from large light cruisers of 22,000, *Furious, Glorious* and *Courageous*, and *Eagle*, which had begun as a battleship of about the same size, but was much slower and carried fewer planes. The planes themselves were notably inferior to those of the US navy or Japanese navy, having been fatally handicapped by the British decision to concentrate on the building of land-based planes for the RAF.

To make matters much worse, two of the seven carriers in service at the outbreak of war, *Courageous* and *Glorious*, had been sunk by June 1940, and only one ship had replaced them, *Illustrious*, about the same size as the *Ark Royal*, but with an armoured flight deck provided by the Czech firm of Skoda and delivered just before the German entry into Czechoslovakia. It was at this time that, with the exception of the unending struggle against the U-boats, the main weight of the naval war shifted to the Mediterranean as Italy entered the war. This was a contingency which had been possible ever since the Abyssinian crisis of the winter of 1935/6, and plans to deal with it had been made. These called for a torpedo attack on the Italian fleet at Taranto by two squadrons of Baffin torpedo aircraft from *Courageous* and *Glorious*. The Baffin was, even by the standards of 1935, a primitive aircraft with a comparatively short radius of action. To provide for this, additional fuel was concentrated at Janina in northern Greece, the Greek government having undertaken to give Britain any assistance required if the imposition of sanctions led to war.

Now, in 1940, the British carrier force in the eastern Mediterranean again numbered two ships, this time *Illustrious* and *Eagle*. However, by then the *Eagle*, having accomplished four years' trials and sixteen years' normal service, was approaching the end of her life, so that a series of near misses by Italian bombers in the early weeks of the war in the Mediterranean had put her out of action in October and November when she was needed for the attack on Taranto. This had again become the most important of the Italian naval bases, situated as it was athwart the British route linking Gibraltar, Malta, Suez, India and the Far East.

In addition to the troubles of *Eagle* various other misadventures delayed the attack, originally fixed for Trafalgar Day and finally postponed, following a hangar fire in *Illustrious,* until 11 November. It was decided that *Eagle* would have to be left behind, but the combined forces of both *Illustrious* and *Eagle,* embarked in the former ship, provided twenty-one Swordfish torpedo bombers.

Confronting them was the Italian fleet with all its 6 battleships, as well as 3 heavy cruisers. Taranto itself was defended by 21 batteries of 3.4-inch anti-aircraft guns and 193 machine-guns, as well as the 92 heavy anti-aircraft guns mounted in the ships. There were 90 barrage balloons provided but they were without the necessary hydrogen, following the loss of many balloons and their hydrogen in a storm, and around the harbour were seven miles of nets. This did not prove to be enough; for the British torpedoes were set to run under the nets which did not stretch from the surface down to the bottom of the harbour but left a gap near the sea bed. These torpedoes were fitted with duplex firing pistols, which exploded either by striking the side of the ship aimed at or by a magnetic device similar to that used in magnetic mines and which exploded under the bottom of the target. These, when they worked satisfactorily, did much more damage than a torpedo which exploded at a ship's side, but they had given some trouble, owing to difficulties from the earth's magnetic field and the rise and fall of the sea, but in their most modern version they were now to be used in action for the first time.

At 1945 *Illustrious* with her escort was steaming 170 miles off the Italian coast. At 2000 eight bells sounded, and the carrier began to vibrate as she worked up to twenty-eight knots, which was nearly her full speed, and turned into the wind to launch her planes.

Faint lights marked the limits of the flight deck on which the first strike of twelve planes was ranged up. From flying control on the island alongside the bridge came a green light and the first of the planes, piloted by Lieutenant-Commander K. Williamson, with

Lieutenant N. J. Scarlett as observer, flashed along the deck, followed one by one by the remaining eleven planes of the first strike. As they disappeared the deck crews began ranging up the nine planes of the second strike to take off half an hour after the first.

Two hours later Taranto began to hear planes. The Italians had no radar but their listening devices were very good, everything possible having been done to develop them, even to the extent of discussing the possible formation of a corps of blind listeners, because it was believed that their hearing was better than that of sighted people.

When the first strike reached Taranto the crew of one plane switched off the Italian radio, over which they had been listening to the performance of an opera, and addressed themselves to the work at hand. The flare-dropping Swordfish began their work, the two planes dropping, one by one, thirty-two in all at fifteen second intervals.

The Italian ships stood out clear and silvery against the background of the slowly descending flares and Williamson picked out as his target one of the two biggest ships in the harbour, the brand new 35,000-ton battleship *Littorio*. He was exactly on cue and, for the first time in history, a torpedo was dropped at night in action. It hit the target, Williamson lifted his plane over the damaged battleship, over a dry dock and then crashed into the harbour. The other planes one by one made their attack. Flak was providing what would have been a magnificent spectacle if anyone had had time to watch, 13,489 rounds of it from the shore alone, with red, blue and orange tracers lit by gun flashes, making a fountain of light, rising and falling. The main concentrations of fire came from land and from a breakwater across the entrance to the harbour, and the curved trajectories of the two sources of tracers met overhead, forming an arched tunnel of multicoloured light, and into this tunnel the Swordfish went, pressing home their attacks. The second strike had arrived by now and the planes were attacking so closely that the splash from the dropping torpedoes drenched them, while the air all around smelled of cordite fumes. Seaplane hangars on the edge of the harbour had also been attacked and were ablaze.

Ashore, 270 miles away in Rome at Supermarina, the supreme naval headquarters, a series of brief signals chronicled the progress of disaster.

First, a series of telephoned air raid warnings which, they could see in Rome, when plotted on the map were heading straight for Taranto. The port was suddenly surrounded by countless flares, the

flash and the sound of bombs dropping, torpedoes exploding and
further details of disaster.

> '*Littorio* hit by three torpedoes.'
> '*Duilio* hit by one torpedo.'
> '*Cavour* hit by one torpedo.'

Those three ships were the three battleships then with the Italian
fleet at Taranto. In addition, a 10,000-ton cruiser, *Trento*, and a
destroyer, *Libeccio*, had been hit by bombs which had failed to
explode.

One by one the British planes returned to *Illustrious*. In the end it
was found that two planes were missing, and all who had returned
safely went down into the wardroom for a supper of bacon and eggs
and a huge cream cake, covered in technicolour icing, the work of
the Maltese Chief Petty Officer Cook. The Commander-in-Chief,
Admiral Sir Andrew Cunningham, signalled '*Illustrious* manoeuvre
well executed' and celebrations were only briefly interrupted by the
news that the attack was to be repeated on the next night. There
were some protests. 'They didn't make the Light Brigade go round
again,' said someone, but the weather shut down and *Illustrious*
did not have to go round again either. Two of the missing men lost
their lives, the other two, Williamson and Scarlett, had been pulled
out of the sea by workmen who set upon them, tearing off their
clothes. They were, however, saved from anything worse by the
arrival of an Italian naval officer, who inquired if they were
wounded and then took them into naval headquarters where they
were given clothing, food and drink, before setting off for a career
of four and a half years in Italian and German prisoner of war
camps.

As for the stricken ships, aerial photographs showed the *Littorio*
and *Duilio* on the bottom in shallow water, run ashore to prevent
their sinking in the middle of the port. The *Littorio* had two holes
in her bottom, one measuring about 48 ft by 32 ft and the other
40 ft by 30 ft. However, loss of life in the sunken ships was small,
twenty-three men from the *Littorio*, sixteen from the *Cavour* and
one from the *Duilio*. The *Littorio* and *Duilio* were eventually raised
and returned to service, thanks to the genius of the Italians in
matters of salvage, but the *Cavour* was doomed, although she survived to be sunk twice more during the war. After Taranto she was
raised and taken to Trieste for repairs, the completion of which were
still six months ahead when she was scuttled at the time of the
Italian armistice in September 1943 to prevent her falling undamaged into the hands of the Germans, who raised her a second

time. She was sunk again by the USAAF in February 1945, raised for a third time after the war was over and finally broken up.

13

Hood *and* Bismarck, 1941

In the whole of naval history there has been nothing to compare with the operations of May 1941, during which the German battle-ship *Bismarck*, then the largest warship in the world, sank the British battle-cruiser *Hood*, which had for more than twenty years itself been the largest warship ever built. Then, after events spread over four days and three thousand miles *Bismarck* was herself sunk by a combined force of 8 battle-cruisers or battleships, 2 aircraft carriers, 11 cruisers, 21 destroyers and 6 submarines with planes flying 300 sorties. To leave nothing undone to catch the *Bismarck* and, if possible, the cruiser *Prinz Eugen* which was with her, these British warships had been called together from all over the North Atlantic, from the limit of the Arctic pack ice off Greenland, in the Denmark Strait, and from Gibraltar and Halifax, Nova Scotia.

The *Bismarck*, built at Hamburg, had moved out of the way of the RAF to the port of Gotenhafen (formerly Gdynia and before that Gdingen) near Danzig in August 1940 and had begun the lengthy and detailed task of 'working up', which meant the testing of the thousands of complicated pieces of machinery and electrical fittings upon which a big ship depends, a process followed by machinery trials, gunnery trials and all the rest at sea. It was a long drawn out and meticulously performed job, which could not be rushed despite the urgency with which Raeder, the C-in-C of the German navy, was demanding its completion, for he had plans for the *Bismarck* as soon as she was ready.

Raeder had been appointed as C-in-C of the navy in 1928, but was obliged to wait until 1935 when Hitler denounced the disarmament clauses of the Treaty of Versailles before he could begin to build up the great fleet with which he intended to challenge Britain. According to Hitler he had until 1948 to prepare, but this date was brought forward to 1945 after Munich. The final plan, called 'Plan Z', was prepared in 1939 and the backbone of it was to be six battleships of 70,000 tons, with eight 16-inch guns – almost twice as large as anything building at that time – together with *Bismarck* and her sister-ship *Tirpitz*, 45,000 tons with eight 15-inch guns, and two smaller battleships, *Scharnhorst* and *Gneisenau*, 31,000 tons with nine 11-inch guns. There were also to be aircraft carriers, cruisers, destroyers and more than 200 submarines. The biggest ships, the 70,000-ton battleships, were to operate against British trade from a base to be established in the Antarctic, while negotiations were planned with Mussolini to permit the use of the port of Kismayu in Italian Somaliland, which had been ceded by Britain to Italy in 1924 – an early indication of the strategic importance today of the Horn of Africa, reflected in the Russian activities in that area during the 1970s and 1980s.

Most of all this was in a theoretical future which never happened, save for the huge U-boat fleet which was built between 1935 and 1945. Of the battleships only the *Scharnhorst* and *Gneisenau* and the *Bismarck* and *Tirpitz* were completed, but Raeder did not abandon his plans for commerce raiding with surface ships.

'In the early morning hours of 4 February 1941,' wrote Captain Bidlingmaier, the German naval historian, 'for the first time in the history of warfare, German battleships broke out into the open waters of the Atlantic, and began thereby a second, expanded phase of the war against seaborne trade.'

These battleships were the *Scharnhorst* and *Gneisenau*, commanded by Vice-Admiral Lütjens, and they were setting out on a cruise which, under the title of Operation Berlin, was to last two months from 22 January 1941 to 22 March 1941, during which they sank twenty-two Allied merchantmen of 115,622 tons.

The cruisers also carried out minor raids, but Raeder's efforts regarding commerce raiding were concentrated on getting the big ships to sea again. He was, however, to be considerably disappointed. In the first place the *Tirpitz* was not ready and, secondly, the *Scharnhorst* was badly in need of repairs to her boilers, while an aircraft of Coastal Command had torpedoed the *Gneisenau* and put her out of action.

But the *Prinz Eugen* was available, and although Lütjens mourned the temporary loss of the *Scharnhorst* and *Gneisenau* he was determined to carry out the planned operation. In theory the *Bismarck* and *Prinz Eugen*, working together, would be able to cope with the recent move by the British for the protection of their most important convoys which were now being escorted by battleships. As a result of this *Scharnhorst* and *Gneisenau*, in the course of their Operation Berlin, had three times come upon British convoys protected by battleships and were obliged to withdraw in accordance with Raeder's orders. Now it was considered that the *Bismarck* was strong enough to attack a British battleship and that

she should do so, while the *Prinz Eugen* would have the task
of rounding up and sinking the merchant ships in the convoy.

On the afternoon of 18 May with her band playing, *Bismarck*
left Gotenhafen and headed westward towards the exit from the
Baltic. A little later she was joined by *Prinz Eugen*, while Lindemann, captain of the *Bismarck*, spoke to his crew over the ship's
public address system, and told them that they were sailing on a
commerce raiding cruise that was to last for three months. On the
morning of 20 May the two German ships, with escorting destroyers,
caught up with, and then passed, the Swedish cruiser *Gotland* off
Marstrand on the Swedish coast. *Gotland* reported to Stockholm
what she had seen, intelligence which, thanks to the efforts of the
British naval attaché and the pro-British feeling amongst most
Scandinavians, was in London a few hours later and then sent to
the *King George V*, flagship of Sir Jack Tovey, commanding the
Home Fleet. The German ships were obviously heading for the
open waters of the Atlantic to reach which they had to choose
one of two routes past Iceland, either that between Iceland and
the Faeroes or between Iceland and Greenland, through the
Denmark Strait.

Both gaps were being patrolled by British cruisers, *Norfolk* and
Suffolk in the Denmark Strait and *Arethusa*, *Manchester* and
Birmingham between Iceland and the Faeroes. Lütjens chose the
Denmark Strait and was quickly picked up by *Norfolk* and *Suffolk*.
The latter was the key ship on the British side in the first phase of
the action, for she carried a new and up-to-date radar, while the
set in the *Norfolk* was primitive and of very little value. When the
British cruisers picked up the enemy they took station astern and
followed them into the Atlantic. There was mist about on the port
side of the cruisers, which gave good cover while, at the same time,
Suffolk kept in touch with the enemy by emerging from time to
time to check their position.

Meanwhile, coming up from the south-east from Scapa, were the
first two of the British heavy ships, the battle-cruiser *Hood*, flying
the flag of Vice-Admiral Lancelot Holland, and the battleship
Prince of Wales, the two escorted by six destroyers. *Suffolk* at this
point lost the Germans, but then found them again.

Captain Roskill wrote in his *The War at Sea*, 'Admiral Wake-
Walker's two cruisers now proceeded to carry out, with great skill
and determination, the traditional role of ships of their class in
touch with a superior enemy squadron. In spite of steaming at high
speed through rain, snow, ice floes and mirage effects they held on,
the *Suffolk* on the enemy's starboard quarter and the *Norfolk* to
port.' [16]

But there was one gap in this part of the shadowing, for shortly after midnight on 24 May *Suffolk* held on too long on one leg of a zigzag and could not pick up the enemy when she resumed the pattern of her pursuit. Accordingly, to search for the enemy Holland sent his six destroyers *Electra, Anthony, Echo, Icarus, Achates* and *Antelope* to the northward, while he took *Hood* and *Prince of Wales* on a southerly course.

An odd feature of the situation of both pairs of big ships at this moment was that, having spent their wartime lives protected by destroyers forming anti-submarine screens around them, both pairs were now alone.

Holland and Lütjens were on converging courses, and at 5.35 am [17] the enemy were in sight from the *Hood*. There followed the last few minutes of the *Hood*'s life, which had lasted twenty-five years, all but one week, for she had been laid down at Clydebank on 31 May 1916, a few hours before the battle of Jutland began. No warship except Nelson's *Victory* was as well known to the ordinary British civilian, and the navy was proud of her, knowing that whenever or wherever she met foreign seamen or foreign ships her looks and size commanded respect and admiration. It is not putting forward an exaggerated claim to say that of all the warships of steam and steel she was the most handsome ever built.

But she had been built a quarter of a century before and it had always been appreciated that her horizontal armour, that is the armour on her decks and turret roofs, was too thin to keep out the heaviest modern shells.

During the inter-war years it had frequently been proposed that the *Hood*'s armour be increased, but government followed government, from that of Lloyd George to that of Neville Chamberlain, without anything being done. Economy had overruled any suggestion for improvement until it was finally agreed that the ship should be rebuilt, but it was then March 1939 and too late for the task to be carried out before the start of the war.

The ship had had a brief, somewhat perfunctory refit in the spring of 1941, but she had not had time to settle down when she was called upon to confront the *Bismarck*. Accompanying her into battle at this moment was the *Prince of Wales*, the most recently completed battleship in the British navy; in fact, so recently completed that she still had on board some of the civilian workmen from Cammell Laird, her builders, putting the finishing touches to her armament.

As the two pairs of ships approached each other on the fateful 24 May, those on board the German ships who could do so were watching the oncoming enemy and, despite the excellence of

German optical work, were unable to make up their minds what ships were facing them. Guesses ranged from destroyers to cruisers, to the *Hood* herself, although this latter guess was very much an outsider. The *Hood* was now steaming almost bows on towards them, shovelling up spray to port and starboard, her bow waves looking like glistening snow drifts. This bows-on approach was a feature of the action which has been much discussed and argued about ever since, and a complete explanation of why it was ordered by Admiral Holland has never been worked out. The obvious disadvantage of the bows-on approach was that only the two foremost of *Hood*'s four turrets would be able to fire on the enemy, while the two after turrets would have to remain idle. On the other hand, there was urgent need to close the range as quickly as possible, for at a longer range the trajectory of shells would mean that they were dropping nearly vertically onto the almost unprotected decks of the British ships but, as the two ships approached each other, the elevation of the guns would be depressed and the trajectory would be almost flat so that the shells would hit the sides of the ship, the most important parts of which were protected by vertical armour twelve inches thick.

Both sides opened fire. The two German ships dropped their salvos right around their targets, throwing up curtains of splashes, with each splash 200 feet or more in height. *Hood* did less well and there was an additional complication in that, firing at the leading German ship, she thought she was engaging the *Bismarck* and that the *Prinz Eugen* was the second ship. In fact matters were the other way round, a confusion that resulted from the fact that the *Bismarck* and *Tirpitz*, together with the heavy cruisers, *Prinz Eugen* and *Admiral Hipper*, looked at a distance almost identical. As a result, the *Hood* opened fire at the *Prinz Eugen*, while the *Prince of Wales* fired at the *Bismarck* and was soon in trouble, for one of her forward 14-inch guns developed a defect after one round. The *Prinz Eugen*, with one of her 8-inch shells, started a fire on *Hood*'s upper deck between her second funnel and the main mast. What was burning were ready-use supplies of a device known as the UP or unrotated projectile, a kind of rocket which, when fired into the air, discharged long wire ropes into which it was intended enemy aircraft would fly. It was never successful and was soon abandoned, but at this moment it had set fire to the British flagship.

Then, at 6 am, just as Holland ordered his two ships to turn to a course more or less parallel with that of the enemy to enable the full broadside of each ship to be used, the *Hood* was hit again by a shell which penetrated the upper deck and is believed to have burst in the 4-inch magazine, the explosion of which then detonated the after 15-inch magazine. People watching the ship saw a column of flame, described as four times the height of the mainmast, followed by a mushroom-shaped cloud of smoke, but oddly no noise. Within two or three minutes dreadful news came to the bridge. The officer of the watch reported that the compass had gone, and then the quartermaster that the steering had gone. The captain ordered that emergency steering be switched on and the ship suddenly turned over.

Three men out of the ship's company of 1,419 survived, a midshipman, a signalman and an able seaman. The first two were on the bridge and the other was on the upper deck.

In the water one of them saw that the ship had broken in two, with the 200-foot bow almost vertical, standing out of the water, as the *Prince of Wales* steamed by engaging the two Germans. Despite the brave face the *Prince of Wales* was putting on matters things were far from well aboard her, for half her 14-inch guns were out of action with their mechanism repeatedly breaking down, despite the endeavours of their crews and of the civilian workmen who had been brought along to finish their mountings.

At 6.00 am the *Hood* had blown up, leaving the Germans free to concentrate on the *Prince of Wales* and at 6.02 she was hit on the bridge by a shell from the *Bismarck* which killed or wounded everybody there except the Captain, the Chief Yeoman and the Navigating Officer, who was wounded. Blood dripped down the voice pipe leading to the plotting office below. Another 15-inch shell hit the *Prince of Wales* as well as two 8-inch from the *Prinz Eugen*, and Captain Leach, commanding the *Prince of Wales*, knowing that Tovey and his battleships and carriers were headed up north towards him decided to break off the action when the after quadruple 14-inch turret also broke down. Accordingly he dropped astern but continued to follow the enemy.

But, despite the troubles which had ruined the performance of the armament of the *Prince of Wales*, she had scored three hits. One of these was to have a vital effect on the result of the action for, although it had passed through the port bow on the water-line and out the other side without exploding, it had nevertheless let seawater into two oil tanks, the contents of which were spilling into the sea and leaving a long trail of oil visible for miles on the surface. This shell had also smashed up suction valves through which oil was supplied to the boilers, with the result that *Bismarck* was deprived of a thousand tons of oil at a time when every ton was going to be needed.

The two groups of ships steamed south at full speed, Lütjens and Brinkmann, captain of the *Prinz Eugen*, wishing to get clear of their trackers, while Rear-Admiral W. F. Wake-Walker, with *Norfolk*, his flagship, and *Suffolk* and the *Prince of Wales* now in company, tried to make sure that they did not. To the south-east, on his way down from the coast of Iceland, Tovey was gathering everything that he could from all over the North Atlantic, and at 4 pm he detached the carrier *Victorious*, with the 2nd Cruiser Squadron (*Galatea, Aurora, Kenya* and *Hermione*), to steer for the *Bismarck* and, when within 120 miles of her, to fly off 825 Squadron of nine Swordfish (torpedo planes), and 802 Squadron of six Fulmars (trackers), to make an attack which might stop her, or at least slow her down. Neither the carrier nor the planes were fully efficient, since there had not been time for a thorough work up, so that once again an emergency had sent ships and planes to sea in a condition far from first class.

The two Squadrons, led by Lieutenant-Commander Eugene Esmonde, had taken off in foul weather and then had difficulty in finding their target. Instead they encountered a United States coastguard cruising cutter, *Modoc*, then on neutrality patrol, which had been ordered to check on belligerent action within 500 miles of the east coast of the American continent.

After this surprise meeting Esmonde and his planes regrouped and this time found the *Bismarck*, now fully alerted, blazing away with her sixty-four guns, ranging in size from 15-inch to 20-mm.

As the British planes came in the noise of gunfire and engines was so great that it was impossible for the man at the wheel of the *Bismarck*, Leading Seaman Hansen, to hear the helm orders, so that he was obliged to use his own judgment and steer the ship as best he could to dodge the torpedoes whose tracks he could see from his position on the bridge. Of the nine torpedoes he attempted to dodge, only one hit the ship, apparently in the area where she was best protected for, although it exploded throwing up a great splash of water, *Bismarck* was undamaged. As for Leading Seaman Hansen, he was awarded an immediate Iron Cross First Class by Lütjens.

This attack by the aircraft from *Victorious* was, somewhat surprisingly, witnessed by the *Modoc*. These American patrols regularly broadcast their positions to avoid incidents, and soon found themselves involved in rescue work when U-boats had made a successful attack on an Allied convoy. Thus, the *Modoc* on 24 May had been looking for survivors from the nine ships which had been sunk during an attack on Convoy HX 126 that had lasted four days.

Three hours later, at 3.06 am, the *Suffolk* lost the *Bismarck*, now steering a course direct for St Nazaire. How Lütjens had been brought to this decision nobody knows. There has been a strong suggestion, although based on nothing more substantial than that Lütjens and Lindemann were seen in deep discussion together, that Lindemann was urging that *Bismarck* should turn back and fall upon the damaged *Prince of Wales*, finish her off and then return triumphantly to Germany, having sunk two of the enemy's biggest battleships.

That this did not happen may have been due to Lütjens' unwillingness to disobey Raeder's order to carry out the three months commerce raiding cruise which had been planned. While this operation, code-named *Rheinübung* (Exercise Rhine) was being discussed early in its planning stage, Lütjens is said to have remarked that both his predecessors as *Flottenchef* (Fleet Commander) had been relieved of their commands by Raeder for refusing to take the offensive against British shipping; the first had been Admiral Boehm, in October 1939, who declined to carry out a minelaying expedition unsupported by any big ships, and the second had been Admiral Marschall, who had been ordered, at the end of the Norwegian campaign in June 1940, to take the three biggest German warships then in service, *Scharnhorst, Gneisenau* and *Admiral Hipper*, and raid out into the oceans. Things, however, went wrong for the Germans from the start, for the *Scharnhorst* and *Gneisenau* were both put out of action by British warships and Marschall refused to carry out the operation with the *Admiral Hipper* alone.

After Marschall, it was the turn of Lütjens and, the *Scharnhorst* and *Gneisenau* having been repaired, between January and March 1941 he carried out the very successful Operation Berlin in which over 100,000 tons of Allied shipping were sunk.

It was accordingly planned to repeat this success in *Rheinübung* with *Bismarck* and *Prinz Eugen*.

The *Spichern* was one of five ships, of which four were oilers, sent to sea in preparation for *Rheinübung* to provide fuel and stores for the *Bismarck* and *Prinz Eugen*, as well as for other raiders and U-boats. During the years 1939–41 fifty-eight German freighters or tankers replenished surface ships and U-boats at sea, a procedure which could well have been copied by the British, especially as war with Japan became more and more likely. As it was, reliance was placed on the chain of British possessions around the world, which looked impressive in red on the map but were, for the most part, too remote or unsuitable. The effect of this on the hunt for the *Bismarck* is shown by the fact that, when *Rheinübung* was approaching its crisis on 25 May and when every ship was needed, both *Prince of*

Wales and *Repulse* had to break off the search and steam half-way across the Atlantic to refuel; the *Prince of Wales* going to Iceland and the *Repulse* to Newfoundland while, on the following day, the *Victorious* too had to make for Iceland to refuel.

Even more important, the British destroyers had to leave the two big ships, *King George V* having now been joined by *Repulse*, which they were protecting just as operations were moving towards that part of the Atlantic within range of the biggest German bombers and in which U-boats were being concentrated to cover the approaches to Brest.

From 3.06 on the morning of 25 May when *Suffolk* had lost contact with *Bismarck*, British and German ships were both steering south-eastward roughly 100 miles apart, with the British on the inside track between the Germans and their French bases, but some 150 miles astern. Had Tovey known that this was the situation he would no doubt have been very reassured but as it was he did not know where the Germans were, nor where they were going. Brest or St Nazaire were always strong possibilities, but so was a break back to Norway or to Germany, passing north around the British Isles.

When the Germans had been at action stations for fifteen hours men of their *B-Dienst*, which listened continuously to British signals, came to the conclusion that the enemy was no longer in touch and parties of men were allowed to leave their stations in turn, to wash and get some hot food. In fact, contact had been lost some time previously, but the *Bismarck*, not realising this and believing that the British knew where she was, had transmitted a long signal which clarified matters considerably for Tovey, as it revealed the German position. But if he knew where *Bismarck* was, he still did not know where she was going.

Meanwhile, hands in the *Bismarck* were set to work building a dummy funnel, in a not very serious attempt to disguise their ship as a two-funnelled British ship of the *King George V* or *Renown* class. Now that the British shadowers had been thrown off the track feeling on board the *Bismarck* was fairly light-hearted and it was suggested that when the dummy funnel was completed smokers would be provided with cigars, pipes and cigarettes and ordered to produce smoke from within.

The uncertainty as to the *Bismarck*'s intentions spread like a ripple from a stone thrown into a pond or rather into the Atlantic Ocean. There was a possibility that she might head for the Mediterranean, and to deal with that the battleship *Nelson*, then at Freetown, was ordered north to Gibraltar, but wherever the *Bismarck* went in the next few days she would still be in the North Atlantic, and further arrangements would have to be made to cope with that situation. For both sides the next move was up to their submarines.

Eight U-boats were called in from their positions in mid-Atlantic to protect the *Bismarck* on her way to one of her French safe havens. One of the Germans, *U 556*, who had fired all her torpedoes, saw the *Renown* and *Ark Royal* both steam by unescorted and not zigzagging, in easy torpedo range. At the same time, six British submarines gathered around the approaches to the Breton ports, waiting for the chance of an attack as the German ship neared safety.

Missing pieces of seapower now began to fit together, forming a whole which was to destroy the *Bismarck* within twenty-four hours. The next arrivals on the huge sea of battle were the five destroyers of the 4th Destroyer Flotilla, commanded by Captain Philip Vian, who had distinguished himself in February 1940 by the rescue of 299 British seaman held prisoner in the German fleet auxiliary *Altmark*. Vian's destroyers, *Cossack, Maori, Sikh, Zulu* and *Piorun* (the latter Polish), had been diverted from the escort of Convoy WS 8 B, 'Winston's Special', one of the through troop convoys to the Middle East.

Finally it was the turn of the air; first came Coastal Command of the RAF with a vital sighting of the *Bismarck*, followed by a second by the Fleet Air Arm from the *Ark Royal*. Coastal's chance came at 10.30 am on the 26th under circumstances which were kept secret for some thirty years and which have only recently been revealed, notably in *Pursuit, the Sinking of the Bismarck* (Ludovic Kennedy, 1974), which gives the most complete account of the *Bismarck* affair so far published.

The Catalina Z of 209 Squadron was a type of long-range flying boat recently supplied to Britain under the terms of the Lend Lease agreement. With these planes had come a number of American naval air officers whose duty it was to teach the RAF how to fly them. One of these officers, incidentally wearing American uniform, was flying in Z/209, the captain of which was Flying Officer Dennis Briggs, but the American's presence on the scene was not revealed until some thirty years after the action.

Now, at 10.30 on the morning of 26 May 1941, the sea was rough and visibility was poor, though not poor enough to hide a suspicious ship at a distance as Z/209 came closer under the clouds, at a height of about 500 feet. To have a good look Briggs took the Catalina up through the clouds and then brought her down again out of them through a gap and there 500 yards away was the *Bismarck* with, so it seemed, every one of her sixty-four guns in action, splinters from which were punching holes in the flying boat's hull. Briggs got off a signal reporting the situation.

For a moment it seemed that the German ship might be able to get away from Tovey with the *King George V* and the *Rodney* which was now coming down to join him. But, at the same time, another British force, this one from Gibraltar, was coming up. This group, Force H commanded by Vice-Admiral Somerville, comprised the battle-cruiser *Renown*, the aircraft carrier *Ark Royal* and the light cruiser *Sheffield*. The *Renown* was no match for the *Bismarck* – six 15-inch guns against eight; 32,000 tons against 45,000 tons, 9-inch armour belt as compared with $12\frac{1}{2}$-inch. There could thus be no question of a confrontation between the two ships, but what might be possible was an attack by *Ark Royal*'s Swordfish which could, perhaps, sink or at least slow her down so that her pursuers might overtake her.

The sea was now getting up, the end of the *Ark*'s flight deck rising and falling more than fifty feet, while there was a wind of 50 mph over the deck. This was all in the very early days of carrier operations and very few people had imagined the possibility of flying off, finding a remote enemy, attacking and then returning safely to the carrier, although on this occasion one at least did not succeed in doing so, for, returning from a reconnaissance, it was struck a grievous blow from beneath by the rising flight deck and smashed. Another, going over the falling bow, fell herself, so that her wheels hit the sea but the pilot pulled her up in time and she went on her way.

The next difficulty was failure of warship recognition, for a Swordfish, coming upon the two-funnelled British cruiser *Sheffield*, took her for the one-funnelled *Bismarck* and attacked, followed by fourteen other Swordfish, most of whom dropped their torpedoes, all of which missed thanks largely to the skilful ship handling by *Sheffield*'s captain and the order which he had given to his guns' crews not to open fire. Then the Swordfish trailed off disconsolately back to the *Ark*, to rearm this time with contact pistols instead of the magnetic ones they had used for the first attack.

Now things were very difficult, for on the next day the *Bismarck* would come under the German fighter cover, while the *King George V* and *Rodney* would both be obliged to turn for home, being on the verge of running out of fuel, as indeed was the *Bismarck*. However, at 8.53 that evening the Swordfish did find the *Bismarck* and attacked through her concentrated flak, scoring two torpedo hits, one of which did no damage and the second dealing the German ship a blow from which she never recovered. Dropped by a plane flying at wave-top height it struck right aft, its explosion flooding the steering compartments and jamming the two rudders so that the ship could no longer be steered and was very soon at the mercy of the *King George V* and *Rodney*, and then helpless, with her fire control destroyed and all four of her twin 15-inch turrets out of action. The ship's hull remained intact so that, for fear of her falling into enemy hands, it was decided that she should be scuttled and, while this was being done, a torpedo from the cruiser *Dorsetshire* struck her. Thus it is not possible to say exactly what caused the *Bismarck* to sink with a loss of 1,977 lives. Only 115 survivors could be rescued, owing to the reported presence of U-boats in the vicinity.

It is interesting to see that although *Bismarck*'s hull remained intact until the very end of the action, her armament, fire control, and electrical installations had all been put out of action. The reason for this appears to have been that her internal wiring was almost all placed above her main armoured deck and was disabled with comparative ease. Much the same state of affairs applied to the *Bismarck*'s sister-ship *Tirpitz* which was disabled by a Home Fleet carrier raid on 3 April 1944, having been hit by fourteen bombs, and having lost 122 men killed and 316 wounded. None of the bombs penetrated her armoured deck and she was repaired within three months, save for the fact that C turret had been jammed by the explosion of a mine placed under her keel by a British midget submarine in Altenfiord on 22 September 1943.

The awe-inspiring events, which included the chasing of the *Bismarck* and the sinking of the *Hood* and the *Bismarck*, together with the very small number of survivors from the big ships, made it impossible until after the end of the war to establish some of the vital developments which had taken place, one of the most important being the arrival on the scene of radar as a fully fledged instrument of war. Although the principle of its working had long been known it was not until the 1930s that it was really developed, almost in parallel by the British, French and United States navies.

The first two sets of radar to be mounted in big ships were those placed aboard the British battle-cruiser *Repulse* and the German 'pocket battleship' *Admiral Graf Spee* in 1936. Neither had a range of more than five or six miles, too short a distance to be of any value.

However, by 1941 the Germans had successfully developed their radar to such a degree that it is believed that the *Bismarck* was able to straddle the *Hood* with her first salvo, if not to hit her. To 'straddle' signified that shells fired together fell on both sides of the target thus indicating that the exact range had been found.

While the Germans were thus successfully developing radar for gunnery, the British had been able to use it for tracking and thus finding and following an invisible target either on the surface of the sea or in the air.

14

'Air Attack on Pearl Harbor. This is No Drill.'

Signal Tower, Pearl Harbor, to C-in-C Pacific Fleet
7.55 am 7 December 1941

On 2 February 1932 the General Conference for the Reduction and Limitation of Armaments met in Geneva after twelve years of preparation, but despite the amount of time thus spent the opening had to be delayed for several hours because the Council of the League of Nations was meeting to discuss what was in fact the first round in the Sino-Japanese wars, which lasted from 1931 to 1945, and which were cosmetically called in their early stages 'the Sino-Japanese Dispute'.

This rapidly escalating war had begun on 19 September 1931 with what was known as the 'Mukden Incident', when the Japanese army secretly blew up a section of the South Manchurian Railway to give an excuse for the occupation, first of Mukden itself, and then of the whole of Manchuria and Jehol. Manchuria was renamed, becoming the Empire of Manchukuo, and a puppet emperor was found in the person of Mr Henry Pu-Yi, the former Emperor of China. The Mukden Incident itself was followed on 28 January by a Japanese landing at Shanghai.

It was clear in Geneva that the Japanese were once more on the march for the first time since they had presented the celebrated Twenty-one Demands to China in 1915.

The writer remembers well, on one of the very early days of the Disarmament Conference of 1932–4 in the Salle des Pas Perdus, speculating with a Soviet Vice-Admiral, retired aged thirty-eight, as to what would happen next and what would have to be done about it. The time was that of the Great Depression which seemed to be more like a bottomless pit into which the Western powers were rapidly sliding. There were two sets of problems: first, what

were the Japanese going to do next and, secondly, what the
Western powers would, or could, do to stop them.

The Russian and I agreed on the need for Anglo-American co-
operation, based on the Philippines and Malaysia but the Russian
went on to stress the need for French co-operation because of the
importance of their bases in Indochina. This was in the early days
of the reign of the aircraft carrier, and possession of air bases ashore
was believed to be essential to provide cover for coastwise shipping
carrying troops and supplies. In the years that followed I half forgot
my talk with the Russian in 1932 for they were times when the
earth shook and worries and crises piled one on another, each half
hiding the preceding one without solving it.

Not until the fall of France and the news of the successive
Japanese demands made on the Vichy authorities in Indochina did
I remember the Soviet admiral.

The demands began on 25 June 1940 when the Japanese insisted
on being allowed to land troops in the ports of northern Indochina,
and thus were able to cut the Haiphong–Kunming Road which, with
the Burma Road, were the only remaining links between China and
the outside world. So successful had the Japanese been in seizing the
Chinese coastal towns, that on 22 September they began an organised occupation and exploitation of French Indochina. As a counter-
stroke President Roosevelt four days later placed an embargo on
the export of iron and steel from the United States to Japan.

In March 1941 a brief and undeclared war broke out between
Thailand and the French forces in Indochina, started by the Thais
to recover territory which they had lost to France in the 1880s. The
Japanese intervened, ordering the French to cede comparatively
small pieces of territory to Thailand and guaranteeing French
possession of the rest of Indochina on condition that Japan was
assured of a monopoly of the local rice crop and the use of the
Saigon airfield which was within bombing range of Singapore.

The Japanese then announced a protectorate over Indochina, to
be held jointly with France and they followed this by the occupation
of the whole colony.

The next move was obviously President Roosevelt's, and he took
it, freezing Japanese assets in the United States – a step also taken
by Britain and the Netherlands government in London. Oil was, of
course, the key to this move; without foreign assets the Japanese
could not buy any and their tankers lay up empty in the ports of
the East Indies.

This was the moment of truth for the Japanese. Admiral Morison,
in volume 3 of his *History of the US Naval Operations in World
War II*, points out that the Japanese Premier, Prince Konoye, and

Mr Suzuki, President of the National Planning Board, did not believe that this new American measure would be fatal to Japan, but General Tojo, who succeeded Konoye as Premier on 16 October 1941, cast his voice decisively for war. In the final discussions in the Japanese cabinet neither Suzuki nor Konoye believed that the oil embargo made war inevitable, but Tojo did.

Admiral Morison comments: 'The point is important, as American pacifists have echoed the Japanese militarists' claim that war was "necessary" when oil was cut off; that it was we who "provoked" Pearl Harbor. Here Japan's civilian premier and No. 1 production planner admit that it was not necessary.'

Earlier Suzuki had assured Konoye that he could increase domestic production of oil and other materials if allotted a small part of what a Pacific war would cost. But, he remarked, 'opening hostilities is a matter of domestic politics'. A remark 'with much meaning' recorded Konoye.

'Of course,' comments Morison, 'that was the nub of it, had been all along. Only by war could Tojo and the *Showa* (an Oriental version of National Socialism) Restoration crowd rivet their control on Japan.'

So the next Japanese moves were made. On 25 November US Army Intelligence reported in Washington that between thirty and fifty troop transports and supply ships had been sighted off Formosa (now Taiwan) steaming south and heading for Malaya and Borneo.

Four days earlier the Japanese First Air Fleet, the main striking force upon which Tokyo was relying for the great blow it was about to strike, had arrived at Tankan Bay, a harbour in the island of Etofuru in the Kuriles, bleak, fogbound and hundreds of miles away from shipping lanes. The extent of the human habitations at this place was a wireless mast, three fishermen's huts and a small pier.

The whole force was commanded by Vice-Admiral Chuichi Nagumo. Immediately under him was the Air Attack Force, with six carriers, *Akagi, Kaga, Shokaku, Zuikaku, Hiryu* and *Soryu*. The screen for these ships was provided by the light cruiser *Abukuma* and nine destroyers, while they were escorted by the Support Force of two battleships, *Hiei* and *Kirishima*, two cruisers, *Tone* and *Chikuma*, and a fleet train of eight tankers and supply ships. In case the weather was too bad to allow for replenishment at sea, thousands of drums of fuel oil were stowed aboard the big ships, in their 'tween-decks and their superstructures.

Here, in the Kuriles, it was safe to let the crews know what they were going to do and their pleasure and excitement was great.

The Japanese First Air Fleet sailed from Tankan Bay on 26

November 'through thick fog and stormy seas' according to a Japanese officer and, despite bad weather, was able to refuel at sea on 3 December (East Longitude date) and then carried on to the launching point ninety miles north of Oahu, the island on which are Pearl Harbor and Honolulu, which they reached at 6 am on 7 December (West Longitude date). The first task was to make sure that the American fleet really was in Pearl. Four float planes were catapulted off and soon reported that the fleet was where it was expected to be. It was Sunday morning and Admiral Kimmel was known to be in the habit of bringing his fleet back to Pearl regularly so that it might spend Sunday in port. All the ships which the Japanese expected to see were there, except for the most important of all, the three aircraft carriers *Saratoga*, *Lexington*, and *Enterprise*. This was a great disappointment to the Air Operations Officer in *Akagi*, who remarked that he would rather sink the carriers than all eight battleships then at Pearl.

Although it was the Japanese aircraft that did all the damage in this raid it was not the aircraft which were the first to attempt an attack, nor the first to be spotted by the Americans.

The attacking planes had been preceded by what was called the 'Advance Expeditionary Force' whose duty it was to carry out reconnaissance and to sink any ships which survived the air attack. This force consisted of twenty-seven submarines, described by Admiral Morison as 'aggressive', eleven of which carried small aircraft and five midget submarines. The latter were launched by the big boats at about midnight on 6/7 December, and it was one of them which was responsible for first alerting the Americans. At 3.42 am, four hours before the start of the air attack, a small converted minesweeper, *Condor*, sighted a periscope where no periscope should have been and reported this to the destroyer *Ward*, which attacked, sank the submarine with depth charges and gunfire and reported accordingly at 6.54, a report which did not reach the Duty Lieutenant-Commander at Pearl until 7.12. However, as early as 5.20 and 5.34 the signals exchanged by *Condor* and *Ward* had been heard by a shore radio station, which had not thought to pass them on, so that news of what was happening did not reach Admiral Kimmel until 7.25. He left for his headquarters at once and was on his way there when the air attack began.

Later, an investigation into the operations of the Advance Expeditionary Force showed that it had achieved nothing but had lost five midget submarines and one large one.

Not only the navy failed in alertness at Pearl that morning: two army privates became famous for the fact that they were exercising with the radar equipment on the northernmost point of Oahu.

They carried on for half an hour longer than they were supposed to, waiting for a delayed breakfast truck, and passing the time by tracking a strange plane, which they reported, only to be told not to bother about it as planes were expected to be arriving from the mainland.

This warning, like others, was disregarded because no one in authority on the American side believed it possible that the Japanese would attack Pearl Harbor, despite the fact that what had been described as a 'War Warning' had been received on 27 November from Washington.

Nevertheless, they did attack and to quote Admiral Morison: 'Half an hour after the battle opened, *Arizona* was a burning wreck, *Oklahoma* had capsized, *West Virginia* had sunk and every other battleship (except *Pennsylvania* in dry dock) had been badly damaged. By 0825 the Japanese had accomplished about 90 per cent of their objective – they had wrecked the Battle Force of the Pacific Fleet.'

Six different boards of inquiry met to consider the disaster at Pearl Harbor, and between them they produced forty volumes of evidence.

Across the harbour on that Sunday morning church bells ashore could be heard in the fleet ringing for eight o'clock Mass. The ships were getting ready for another day of peace and the last few minutes of preparation for the hoisting of colours. The big ships were lying mostly in pairs off Ford Island in the centre of the harbour, with the fleet flagship, *California*, nearest to the exit. Then behind came an oiler, *Neosho,* and then two battleships side by side, *Maryland* and *Oklahoma*, the outboard ship, then another pair, *Tennessee* and *West Virginia*, followed by *Arizona* with the repair ship *Vestal* alongside, then finally, by herself, the *Nevada.* In addition to the battleships and the *Vestal* there were in Pearl at that time 2 heavy cruisers, 6 light cruisers, 29 destroyers, 5 submarines, 1 gunboat, 9 minelayers, 10 minesweepers and 23 auxiliaries.

A single bomb, dropped on the ramp of the seaplane station, was taken for a few minutes as an accident; then the Japanese red disc (nicknamed 'meat ball' by the Americans) was spotted, fire was opened from all over the fleet and the 'Val' dive bombers got to work, together with torpedo planes.

Kimmel, of course, immediately signalled Washington and when the message arrived, Col. Frank Knox, the Secretary of the Navy, exclaimed: 'My God! This can't be true, this must mean the Philippines!'

In the twenty-five minutes of the attack on the fleet twenty-one

'Kates' each dropped a single torpedo, these attacks being followed up immediately by dive bombers, machine-gunning the crews of the anti-aircraft guns, mostly in exposed positions on the upper decks of the big ships.

West Virginia was hit by six or seven torpedoes on her port side, and by two bombs as well. Her officer of the deck saw the explosion of the bomb on the seaplane station and at once ordered: 'Away Fire and Rescue Party!' Obeying the order many men saved their lives; the torpedoes knocked out all power, light and communications, and the battleship's captain was mortally wounded, while the ship took on a heavy list, corrected by counter-flooding. This prevented her from capsizing, and she settled down on an even keel, on fire from her bow to No. 1 turret, with flames shooting up to the foretop.

Alongside her was the *Tennessee*, which had been protected by the *West Virginia* from torpedoes and which was damaged for the most part only by debris from the *West Virginia* and by flaming oil drifting on the waters of the harbour. The crew of *Tennessee* were able to take care of their own ship and also helped to look after the *West Virginia*, fighting fires all that day and all the next night. On 20 December, with the *Maryland* and *Pennsylania*, she was able to leave Pearl and all three ships sailed for Bremerton, on Puget Sound, which was to specialise in the repair of damaged battleships.

The *Arizona* was the next astern of the *Tennessee*, and was the ship which suffered most grievously in the whole attack. The *Vestal*, lying outboard of her, should have protected her, but the repair ship was too short and of too shallow draft, so that a torpedo passed right under her and struck the *Arizona* beneath No. 1 turret. A bomb then hit the deck alongside No. 2 turret, penetrated the deck and exploded in a forward magazine which had not been flooded because the keys to the flooding valves could not be found. This explosion sent flames 500 feet into the air and destroyed the whole forepart of the ship, killing her admiral and her captain.

Another bomb went down her funnel, one hit the front of No. 4 turret, one hit the boat deck and four more burst on her superstructure between the bridge and the tripod mast. The effect of all this was to blow the ship in half, the after portion resting upright on the bottom while the riven forward half of the ship, with the foremast, turned over on its side and the mast collapsed. Hundreds of men died in these moments as fire spread through the ship, although the anti-aircraft gun crews remained in action until the flames drove them from their guns. In the end, however, 47 officers and 1,056 men were killed, while 5 officers and 39 men were wounded, out of a total complement of 100 officers and 1,411 men.

The last ship in the line was the *Nevada*, hit by only one torpedo, which struck forward and blew a hole measuring 45 feet by 30 feet in her side. She was then dive-bombed, being hit five times. Her senior officer on board, a Naval Reserve two-and-a-half striper, decided at first to get her out to sea and then changed his mind and decided to run her aground, rather than take the risk of her sinking in the entrance to the harbour and bottling up the whole fleet.

As the *Nevada* steamed down Battleship Row she passed the *Maryland* and *Oklahoma*. The latter had been hit almost immediately by three torpedoes and began to turn over, finally lying on her side with part of her bottom out of the water, looking like a basking whale. Boats from the shore put out at once, and men began walking over the hull, banging on the plating and listening for answering sounds that would reveal the presence within the sunken ship of men still alive.

The only battleship not in Battleship Row at the time of the attack was the *Pennsylvania*, in the dry dock which she was sharing with the destroyers *Cassin* and *Downes*. When the attack began the *Pennsylvania* manned her anti-aircraft armament promptly, but the destroyers had trouble, for parts of their main armament had been removed for modification and it was necessary to send members of the two crews to the ordnance depot to get back the missing parts. This was done but as the guns' crews worked to reassemble their weapons low-flying Japanese planes machine-gunned the dry dock and a bomb cut the cables which supplied power from the yard.

The bombers which comprised the second wave of the attack swept over the dockyard at 8.40. They started a fire in the dock and the *Pennsylvania*'s captain ordered it to be flooded. This raised the level of the water in the dock, but it also raised the level of the flames on the water's surface.

The *Pennsylvania* was extricated without serious damage from the blazing waters, but the *Cassin* and the *Downes* had to be written off, although their machinery and various special fittings were removed and rebuilt in other destroyers under construction in the United States.

With all possible speed ships which could move were got under way, sometimes with a certain amount of confusion. Thus, to unmoor the *Nevada*, Chief Boatswain E. J. Hill clambered onto the mooring quay, which was then being shot up by attacking Japanese planes, cast off the lines and then, as the battleship moved away from the quay, jumped into the sea and swam after her, catching her up.

Another attempt to catch up with his ship, temporarily less

successful, was made by one of the destroyer captains, who dashed
down to the harbour, only to see his ship and the rest of his squadron moving out to the open sea. He procured a motor boat and set
off in pursuit, but his squadron commander would not stop and he
was obliged to abandon the chase, leaving his ship in charge of the
senior officer on board, an Ensign of the USNR, a position equiva-
lent, in the British navy of those days, to Sub-Lieutenant RNVR,
who was afterwards commended for his conduct in command, a
spell which lasted for thirty-three hours.

In all, 183 Japanese planes took part in the attack. First came 40
Kate torpedo planes with special shallow running torpedoes, then
49 high level bombers, also Kates, 51 Val dive-bombers and 43
Zeke fighters, to hold the ring and stand off American fighters
should they be able to intervene, although this was considered unlikely as plans for the attack included strikes against the army airfields ashore. These were extremely successful, since the army was
expecting only attacks by saboteurs and had, accordingly, concentrated planes in easily defensible areas which were splendid targets,
with the result that the US army and navy, between them, lost 248
planes on this day, while the Japanese lost only 29 in action, together
with a small number which crashed when landing back on their
carriers.

American casualties totalled 2,403 killed and 1,178 wounded.
Their monument is the wreck of the *Arizona*, its place in the
harbour marked by a kind of island built over the battleship's hull,
flying the Stars and Stripes and guarded by a dozen men whose duty
it is to return the honours paid to the sunken ship by passing
vessels.

As for the Japanese, they had lost fifty-five men killed.

It had been a terrific victory for them, bought at an almost
infinitesimal cost, but it can now be seen as a battle that it had been
unnecessary to fight.

The best argument for not fighting the battle had, somewhat
oddly, been put forward by an American staff officer, Captain
Vincent R. Murphy, USN, giving evidence in 1944 at one of the US
official inquiries. He said:

> I thought that it would be utterly stupid for the Japanese to
> attack the United States at Pearl Harbor . . . I thought that
> the Japanese could probably have gone into Thailand and
> Malaya, and even the Dutch East Indies . . . I did not think
> that they would attack at Pearl Harbor, from my point of view.
> We could not have materially affected their control of the

waters that they wanted to control, whether or not the battle-ships were sunk at Pearl Harbor. In other words I did not believe that we could move the United States Fleet to the Western Pacific until such time as auxiliaries were available, as the material condition of the ships were improved [sic], especially with regard to anti-aircraft and until such time as the Pacific Fleet was materially re-inforced. I thought it would be suicide for us to attempt, with an inferior fleet to move into the Western Pacific.

Admiral Morison comments: 'And so it would have been, as the fate of *Prince of Wales* and *Repulse* indicates.'

15

Prince of Wales *and* Repulse, *1941*

The attack on Pearl Harbor occurred on the same day as the other Japanese attacks on Thailand, Malaya, Borneo and Luzon in the Philippines, and all were part of a great design by Tokyo to secure oil, rubber and tin. Once in possession of these supplies the Japanese government believed that it could hold off any Anglo-American counter-attack and proceed with the organisation of their vast new Empire, which already included a large part of China and which would be so strong that, in the long run, the British and the Americans would weary of counter-attacking and would leave them in possession of a vast proportion of the world's surface and of its resources.

From 1931 onwards this seemed more and more the shape of things to come, especially as the West was by then in the grip of the Great Depression. However, bit by bit, they did begin to emerge to find themselves faced not only with the Far Eastern situation and the Japanese carrying out their plans, but also with Hitler carrying out his.

France left the war and Russia entered it; the American attitude was unclear to all, including the Americans themselves; and there seemed no definite prospect of Britain being able to rely on any outside help for the defence of her possessions in the Far East or of Australia and New Zealand. The Japanese occupation of the airfields around Saigon and the priceless harbour of Camranh Bay emphasised that there was not much time left to prepare for an all-out struggle, and preparations for the dispatch of a British Fleet to the Far East had to be made during the autumn of 1941. The backbone of this fleet would have to consist of capital ships but of these there were only two thoroughly up to date, *King George V* and *Prince of Wales*, who were with the Home Fleet mostly based in Scottish waters. From there they had been able to deal with the *Bismarck* and were now facing *Bismarck*'s sister-ship, the *Tirpitz*, just on the point of finishing her work-up and about to leave the Baltic for a base in Norway.

In the Mediterranean, based on Alexandria, were *Queen Elizabeth*, *Valiant* and *Barham*, and at Gibraltar, *Nelson* and *Renown*. Nearly all these ships had been working hard since the beginning of the war two years previously, so it is not surprising that no fewer than six out of a total of fifteen were refitting at this time, *Malaya*, *Repulse* and *Royal Sovereign* at home, and *Rodney*, *Warspite* and *Resolution* in American dockyards. Two more ships, *Ramillies* and *Revenge*, were escorting convoys in the North Atlantic in case of a sortie by *Tirpitz*, or by *Scharnhorst* and *Gneisenau* in Brest.

At this stage the Admiralty believed that it could assemble seven capital ships for an Eastern Fleet, *Nelson*, *Rodney*, *Renown*, *Ramillies*, *Resolution*, *Royal Sovereign* and *Revenge*. *Nelson* and *Rodney* were middle-aged ships, heavily armed and armoured but slow, *Renown* (a sister-ship of the *Repulse*) was fast but with armour of only medium thickness (originally manufactured before World War I for a Chilean battleship taken over by the British navy and converted into the aircraft carrier *Eagle*). With the capital ships were to go to the Eastern Fleet one aircraft carrier, ten cruisers and twenty-four destroyers. The whole force would take nine months to assemble, in comparison with the seventy days that had been considered the time needed before the start of the war in Europe. To wait nine months, through the winter of 1941–2 with the war in a critical state in North Africa, Russia and the Far East, was not to be thought of and something had to be done at once with whatever ships could be collected.

British capital ships and carriers at this time made a very small pack, which everyone tried to shuffle in different ways; the Prime Minister, the Admiralty and the Foreign Office, each with plans of their own, while appeals for help from the authorities of India, Australia and New Zealand were all coming in, but until Pearl Harbor had actually taken place, there could be no certainty that the United States would enter the war.

It was decided on 20 October that the *Prince of Wales* should go to the Far East, picking up the *Repulse* on the way at Cape Town, where she would arrive with a convoy on its way to Suez. Rear-Admiral Sir Tom Phillips was appointed in command, with the acting rank of Admiral. Phillips had spent the first years of the war as Deputy, and later Vice Chief of the Naval Staff. Small and intense, he was no believer in the power of aircraft to sink capital ships if they were well protected and had sea-room in which to manoeuvre – a belief which seemed well-supported by the events of the first two years of the war.

Meanwhile between 27 September and 18 December disaster followed disaster for the British big ships. On 27 September *Nelson* was torpedoed and put out of action by an Italian bomber and was not ready again until April 1942. On 3 November the brand new carrier, *Indomitable*, had run aground in the entry to the harbour of Kingston, Jamaica. The ship was intended to work with the *Prince of Wales* and *Repulse* but this accident put her out of action for several weeks. On 14 November the *Ark Royal* was sunk by *U 81* while *U 331*, on 25 November, sank the battleship *Barham*.

This was the situation when the *Prince of Wales* and *Repulse* were sunk by Japanese aircraft on 10 December and the two sister-ships, *Queen Elizabeth* and *Valiant*, were sunk by Italian frogmen in very shallow water and on an even keel, so that the enemy never learned what had happened until the information was too late to be of any use to them.

The Commanding Officer of the Italian frogmen was Prince Valerio Borghese who after the war played a leading role in Italian right-wing politics until his death in 1974. The leader of the party which sank the two British battleships was Lieutenant Durand della Penne, who was awarded the Medaglia d'Oro, Italy's highest decoration. The medal was actually presented to him by Rear-Admiral Vaughan Morgan, on board the *Valiant*. Admiral Morgan had been Captain of the *Valiant* the night she had been sunk, Italy in the meantime having come over to the side of the Allies.

This long series of British disasters meant that the *Prince of Wales* and *Repulse* were on their own when they reached *Singapore*, save for the destroyers *Electra*, *Tenedos*, *Express* and HMAS *Vampire*.

While the two big British ships were on their way to the Far East it had been necessary to decide how best they were to be used. There were two different ideas, both of which were concerned, primarily, with covering the stream of ships taking men and stores to the Middle East via the Cape of Good Hope. The close cover for these convoys was to be provided by old 'R' class battleships of pre-1914 design, while the *Prince of Wales* and *Repulse* were to be based either on Trincomalee in Ceylon or on Singapore. If the ships were based on Ceylon their task would be defensive, protecting Allied convoys to the Middle East. This idea did not satisfy Churchill, who had given much thought to the problems which the big German surface warships presented if used as raiders, and particularly the *Tirpitz*, the most powerful of them all, and he

believed that it might be possible for the *Prince of Wales* and
Repulse to act in the same way, operating from concealed bases in
the islands of the Western Pacific or the Indian Ocean.

The *Prince of Wales* sailed from the Clyde for Singapore on 25
October with an escort of four destroyers. The composition of this
escort changed several times during the voyage, as destroyers of any
type were hard to come by and Phillips had to be satisfied with any-
thing he could pick up on his way east. He therefore ended up at
Singapore with four ships of three different types, *Express, Electra,
Vampire* and *Tenedos*, ranging in age from six years (*Express* and
Electra) to twenty-three years (*Tenedos*) and twenty-four years
(*Vampire*) – the latter Australian.

When the *Prince of Wales* arrived at Cape Town Phillips flew up
to Pretoria for a meeting with the South African Premier, Field
Marshal Smuts, who pointed out that current planning called for
the British and American fleets to be stationed in two groups, the
British at Singapore and the Americans at Pearl Harbor, separated
by a distance of 5,000 miles, and both forces inferior in size to the
fleet of Japan which stood between them. This, Smuts said, laid
both the Anglo-Saxon fleets open to a terrible defeat if the
Japanese turned with their combined force against first one,
and then the other. In fact, however, the Japanese did not do
this but instead struck separately and almost simultaneously at both
fleets.

On their way through the tropics the crews of the two big British
ships learned, if they had not known it before, that their ships had
been designed primarily for service in the North Sea and North
Atlantic. With only primitive air conditioning, and temperatures
below decks ranging from 95 degrees on the mess decks to 105 to
136 degrees in engine and boiler rooms, the crews of all the ships
were under a great strain, 60 per cent of whom were under twenty-
one years old.

The *Prince of Wales* and *Repulse* met off Ceylon and, after they
had sailed for Singapore, all six ships officially became Force Z, the
Prince of Wales and the destroyers having previously been known
as Force G. The newly named Force Z arrived at Singapore
on 2 December, and at once Phillips in Singapore and Churchill
in London began to make plans for it to leave as soon as
possible.

Originally the dispatch of the ships had been intended as a threat
which would deter the Japanese from their grandiose plans for the
conquest of the Far Eastern possessions of Britain, the United States
and the Netherlands. Any possibility that this would, in fact, deter
them disappeared on 6 December when Phillips, who had flown to

Manila, was in conference with his American opposite number, Admiral Thomas C. Hart, commanding the US Asiatic Fleet. The talks were interrupted by an American officer who brought a report by the pilot of an Australian Hudson that a large Japanese convoy, which had last been reported in Camranh Bay, was now at sea, heading either for Siam or for Malaya.

Phillips had to concentrate his ships at once and fly back to Singapore. The *Repulse*, with *Tenedos* and *Vampire*, was on its way to Darwin to show the flag, having embarked 10,000 bottles of beer for that purpose. The Japanese, meanwhile, had begun laying 1,000 mines in a great field stretching due east from the Malayan coast to catch any British ships which might steam northward from Singapore to interfere with any Japanese landings.

The war between Japan and Britain began at 0045 Singapore time on 9 December, which was seventy-five minutes before the attack on Pearl Harbor. In moonlight and at high tide Japanese troops landed to capture the airfield at Kota Bahru. On the evening of the same day Churchill wrote: 'The ships must go to sea and vanish among the innumerable islands.'

However, as Captain Roskill points out in *The War at Sea*: 'By then it was too late to implement this strategy, for the squadron was already at sea, seeking the enemy.'

Martin Middlebrook and Patrick Mahoney in *Battleship*, which is the most complete account of the end of the *Prince of Wales* and *Repulse* so far published, say: 'When the Admiralty in London had heard of the sighting of the Japanese convoys at sea two days earlier, a signal had been sent to Admiral Phillips asking what action he proposed to take.' Lapsing into a colloquialism of a later date, the authors comment: 'It was a good question. The sending of the two capital ships to Singapore had always been intended as a political deterrent and no detailed planning had taken place about what was to be done if that deterrent failed.'

Bad news, consequent upon that failure, at once began to flood into headquarters at Singapore. The Australian Hudsons and the shore defences at Kota Bahru had damaged all three of the Japanese transports which were landing troops, but the RAF ground personnel abandoned their positions without orders and, after setting fire to the base, disappeared by lorry into the night. In addition to Kota Bahru, Singora and Sunghi Patani also fell to the first wave of the Japanese landings which, according to local experts, should not have been possible in the face of the north-east monsoon.

The target immediately available against which Force Z could strike was the Japanese transports off the coast of Malaya and in the

Gulf of Siam. The American command in Honolulu had an excellent code-breaking unit, the result of whose work was transmitted to the British at Singapore. From this it was learned that, apart from the invasion convoys, there were two Japanese covering forces, the close cover being provided by four 10,000-ton cruisers, *Mogami, Mikuma, Kumano* and *Suzuya*, and the distant cover by the battleships, *Kongo* and *Haruna*, both groups having escorting destroyers in company.

In addition to the decrypts the other intelligence source, if available, was air reconnaissance and application was made to the RAF for help. The senior RAF officer, Air Vice-Marshal Pulford, was asked both for reconnaissance a hundred miles north of Force Z at daylight on 9 December and reconnaissance off Singora at first light on 10 December, together with fighter cover off Singora during daylight on 10 December.

Pulford's answer came back. He could provide the reconnaissance for first light on the 9th but regretted that the fighter cover requested would be impossible.

At the time of the Japanese attack the defences of Malaya and Singapore had, between them, only 158 operational aircraft, 'the best of them being a slow American fighter, the Brewster Buffalo (F2A), and the Bristol Blenheim bomber,' state the authors of *Battleship*.

> Export models sold to Britain and the Netherlands were stationed in the Far East when the Pacific war broke out; they were slaughtered by the more agile Japanese A 6 (Zero) and Ki 43 (Hayashira); the F2As fared no better against the Zero at Midway; most of the Marine F2As based there were lost in action against the attacking Japanese fighters. A disgusted Marine officer afterwards wrote that to order F2A pilots against Zeros was to order them to commit suicide.[18]

The same superiority of the Japanese fighters over the Buffaloes at that time was to be exemplified, within a few hours, by the Japanese torpedo bombers in comparison with their British counterpart, the Swordfish.

In addition to the failure to judge the material quality of the Japanese naval air service correctly, there was another misjudgment equally as faulty and equally disastrous, namely the lack of a reliable appreciation of the situation in which the two great British ships would be placed.

Somehow or other, the belief amongst the British was widespread that the Japanese brain was incapable of matching the Western brain. All sorts of curious misapprehensions flowed from this belief, as reported by Professor Arthur Marder in his book *Old Friends New Enemies.*

One of the few British observers who appreciated accurately elements of Japanese military strength was an attaché, a soldier this time, whose dispatches were ill-received at the War Office, where it was claimed that they exaggerated the quality of Japanese material and training. The accuracy of these forecasts was not confirmed until the fall of Singapore.

These items, and many similar, made a fog of war into which Force Z, ill-equipped and ill-informed, sailed when it left Singapore just after five o'clock on the afternoon of 8 December. The tropical twilight soon closed down as the ships steered north-east and then north, outside the Anamba Islands and the eastern end of the Japanese minefield of a thousand mines. Strict radio silence was ordered for the Force but they could, of course, receive signals from other ships and from shore, including one from RAF Singapore saying that although it might be possible to supply some reconnaissance, standing air cover would not be available on 10 December, when the Force was to turn west and make a high speed run to attack Japanese invasion groups. There would not be enough fighters to keep even a small force continuously airborne over Force Z, but it might be possible to summon aid from Singapore in the event of a Japanese attack.

With this in mind one Buffalo Squadron, No. 453 (Australian), had been allocated to support Force Z on call, but the success of its intervention would depend upon how far north the British ships had gone before they were attacked.

Shortly after dawn on 9 December the British ships had passed the Anambas. There had been no thunderous dawn, the weather continued bad and favourable to the British purpose of a secret approach, for it was misty with intermittent rain squalls. Then, at about 1330, a patrolling Japanese submarine *I 65* caught sight of the big British ships through its periscope and began to follow them on the surface, until she had to dive when she sighted a seaplane, mistakenly thought to be British but actually from one of the Japanese light cruisers. The joint efforts of the communicators of *I 65*, *Kinu* and Vice-Admiral Ozawa who, in his flagship *Chokai*, commanded the Close Cover Force, failed at first to get the signal through. When this was at last successfully done Ozawa began to collect his planes and his ships to attack the British force, which he had believed still in Singapore. Back at Saigon planes armed with bombs for a raid on Singapore were hastily rearmed, under the sweltering heat of a tropic night, with torpedoes for an attack on the British fleet. A fact that might have been of the greatest importance was that only one torpedo per plane was available in the Japanese advanced bases around Saigon.

Meanwhile Phillips was making his plans for the next day, which was to be the fatal 10 December. After the action was over Captain L. H. Bell, RN, a member of Phillips' staff, prepared a memorandum in which he laid down what he believed to have been his admiral's plan. In a document now in the Public Record Office (ADM 199/1149) he wrote: 'The Admiral's plan had been to detach the destroyers at midnight 9/10th and make a high speed descent on Singora with the less vulnerable *Prince of Wales* and *Repulse* . . .' but suddenly less than two hours before sunset the weather, which had continued persistently misty and most favourable to the British, changed. All clouds were swept away and instead there was brilliant sunshine and three Japanese reconnaissance seaplanes.

Half an hour after sunset *Tenedos*, the smallest of the British destroyers and the one with the least fuel capacity, was on the point of returning to Singapore to refuel while Force Z itself turned inward towards what was reported to be the invasion beach, working up to twenty-six knots. Then *Electra* sighted a flare about five miles ahead and Phillips ordered his ships to turn away. Later, much later, when the Japanese records could be investigated, it was seen that the Japanese plane had dropped its flare by mistake right on top of their own cruiser *Chokai*, which was in company with her four sister-ships. Had Force Z headed towards the flare, rather than turned away from it, the ten 14-inch and six 15-inch guns of the British battleships could, in all probability, have secured a very real success against the Japanese ships which, like the 10,000-ton cruisers of all navies, had only the very minimum of armour.

But Phillips knew that he had been seen, and could not sail into great danger on the off-chance of meeting a group of weaker ships. Accordingly, he held on for Singapore only to be diverted by another signal, reporting a Japanese landing at Kuantan, further down the coast towards Singapore. This was a very important strategic location, for a landing there would cut the communications along the east coast of Malaya, between the British and Indian troops fighting in the north and their base at Singapore. In addition, from Kuantan a road ran right across the peninsula at a point 180 miles from Singapore. Phillips ordered his ships to increase speed and to make for Kuantan.

Ever since he had left Singapore he had maintained rigorous radio silence, so that headquarters in Singapore knew nothing although it was here that Rear-Admiral A. F. E. Palliser, Phillips' Chief of Staff, had been left behind to link the ships and the Admiralty in London.

It has been suggested that Phillips believed that Palliser, who had radioed to him the news of the reported landing at Kuantan, would order the fighters of 453 Squadron to give the *Prince of Wales* and *Repulse* cover at that place, but this did not happen and it was not until 1158 that *Repulse* reported to Singapore that the two ships were under air attack. By this time the battle between the British battleships and the Japanese planes was nearly over.

The first indication that things were going very wrong for the British was a signal at 0950 from the detached *Tenedos* that she was being bombed at a position 140 miles south-east of where the battleships were at that time. It was a very unpleasant shock for the British to realise that the Japanese bombers and torpedo planes had so large a radius of action (at least 700 miles), for this meant that Force Z itself might be under attack at any moment.

The Japanese naval air commander at Saigon, Rear-Admiral Matsunaga, had ordered his force of ninety-nine bombers and torpedo planes to prepare for take-off, although they had only just landed from another operation. Nevertheless, 94 out of a total of 99 planes were quickly ready for action. First to take off were to be nine reconnaissance planes, with full tanks and only two 100-lb bombs apiece. During this operation the Japanese planes, Bettys and Nells, the former being twin-engined land based torpedo planes and the latter twin-engined land based medium bombers, were to be stretched to the full. None would be able to carry a full load of fuel and a full load of missiles. Everything would depend upon the reconnaissance planes being able to find the British ships quickly and then guide the bombers and torpedo planes to them in the shortest possible time. The first wave of attackers would be composed of bombers whose missiles, it was hoped, would tear up the superstructures of the battleships, put their anti-aircraft armament out of action and cause serious casualties amongst the guns' crews. Ideally, armour-piercing bombs should have been used for this work but all available bombs of this kind had been sent to the carriers for use in the raid on Pearl Harbor.

The Japanese pilots took off from the bases around Saigon in groups of between seven and nine planes between 5 and 8 am, each man drawing his ration of coffee syrup in a vacuum flask and rice cakes covered with bean paste.

To fly the planes to Singapore and back their engines had to be carefully nursed, a procedure which, though harmful, was necessary and had often been practised by the 22nd Air Flotilla specially chosen and trained for long range flights over the sea.

The first of the Japanese reconnaissance planes reached a position off Singapore, from which they were obliged to turn back with their fuel running low. As they turned back for Saigon one of them saw the British ships ahead. Because of failures in radio procedure the various groups of aircraft failed to carry out planned combined attacks, but came straggling into action one by one. The first to engage were eight bombers at 11.00 am, as the British hoisted large White Ensigns at their mastheads, the traditional indication that a British warship is going into action.

Flying through the enemy flak seven Japanese planes circled round to come out of the sun and then dropped their bombs, only one of which hit, punching a hole in the roof of the hangar of the *Repulse* and bursting on the armoured deck below in the Marines' mess. There was an explosion, smoke drifted from the hole in the deck, and there were some casualties, including a number of men badly burned, who climbed to safety up the inside of a ventilator. A fire party arrived and within a few minutes the fire was put out, while the Walrus pilot was engaged in getting his badly damaged plane over the side before the fire in the hangar below caused the petrol in its tanks to explode. There was then a brief interval. According to plan, the bombers were to be followed by low-level torpedo attacks, while the British anti-aircraft guns were still at maximum elevation, but this did not happen and by the time that the torpedo bombers arrived the *Prince of Wales* and *Repulse* had had time to prepare to meet the nine torpedo planes sweeping in line abreast low over the sea, and bugles and orders over the Tannoy announced the second round of the fight.

Meanwhile, Admiral Phillips had realised that he had made a mistake in trying to control the avoiding action of both the big ships together, rather than let each ship look after herself. It will be remembered that Admiral Holland had tried to control the *Hood* and *Prince of Wales* in the same way in their action with the *Bismarck*.

The spectacle of the Japanese torpedo attack was a startling one to the British, for the enemy came in at an altitude that led many to believe there was going to be a low-level bombing attack, a belief shared by Admiral Phillips. The standard British torpedo plane was the ancient Swordfish or 'String Bag', to which the British navy was attached with that feeling which so often links the British fighting man to defective equipment of which he has to make the best possible use.

Now the Japanese were coming in at 180 mph, and they dropped their torpedoes from 50 to 150 feet, while the usual operational procedure for the Swordfish was to approach at 100 mph and drop from 50 feet.

The Japanese torpedo planes were late, but when they arrived they began the next stage in the destruction of the British ships. They orbited the big ships, took cover in a cloud and then emerged, tearing down in a shallow dive. It was now 1142. Their gun crews alerted by voice and by bugle calls, both ships opened fire with their complete anti-aircraft armaments and the sky seemed covered with little black and grey woolly clouds. Two enemy planes were shot down from a very low height and the New Zealand Walrus pilot, still trying to get his plane over the side, fired at the enemy with his pistol while, as they passed overhead, the Japanese rear gunners fired on the anti-aircraft guns' crews in the open on the upper deck with a mixture of explosive sounds that almost removed the power of thought.

Nine Japanese torpedo planes, stretched out in a rough line abreast, dropped their torpedoes which tore across the dark blue sea, their tracks marked by long white furrows. It seemed for a few moments that the *Prince of Wales* would dodge them all and eight had missed when the ship appeared to stop. A great pillar of water about 200 feet tall rose up and then fell back with a crash. The *Prince of Wales* at once listed to port and began to slow down, her speed dropping from 25 knots to 15 as she settled by the stern. Calls from various parts of the ship reached the bridge announcing misfortunes and disasters, but much worse was the fact that many positions in the ship did not answer at all, and it was discovered that about half the ship's supply of electricity had failed. Middlebrook and Mahoney in *Battleship* comment: 'A warship lives on electricity, and exactly half the *Prince of Wales* was dead.'

In the after half of the ship there was no light, no ventilation, no power for the guns nor for the ammunition supply.

More bad news came in, its full meaning only realised when it was possible to survey the damage done by the Japanese torpedoes (there had been a second one almost simultaneous with the first which had caused little damage). The first hit, however, had settled the ship's fate in an extraordinary and complicated way. Between the ship's hull and the propellers each shaft was held in position by a projection from the ship's side called the A bracket. This had been hit and damaged by the first torpedo, so that the outermost of the propeller shafts on the port side of the ship was no longer held rigid, and began to thrash about. Two hundred and forty feet long, this huge piece of steel was able to enlarge the aperture in the ship's stern through which it passed, as well as similar apertures in bulk-heads within the ship, and the sea flooded in through the outboard shaft tunnel, thence into the port outermost of the four engine rooms and into the after 5.25-inch magazine.

When the lull in the action began at noon the first concern of Captain W. G. Tennant, commanding *Repulse*, was to discover what had been done by his C-in-C to inform Singapore. From this it would be possible to calculate when the Buffalo fighters might be expected to arrive from Singapore, exactly one hour's flying time away. He was amazed to learn from his signal office that the *Prince of Wales* had made no signals at all, so that the Buffaloes were still on the ground, at Sembawang, on Singapore Island. There were to be no survivors from the flagship who could explain why Phillips had continued to maintain wireless silence after the action had started, and it was clear to the enemy where the British force was and of what it was comprised.

Anyhow, at 1158 *Repulse* signalled: 'Enemy aircraft bombing' and gave her position.

Then, at 1220, from the *Prince of Wales*: 'Emergency. Have been struck by a torpedo on port side', followed by a position and then, 'Four torpedoes. Send destroyers.'

This series closed at 1318 with a signal from the destroyer *Electra* which, with the sinking of the two battleships, was now senior officer:

'*Prince of Wales* sunk.'

The lull at noon was ended after twenty minutes when twenty-six Betty torpedo bombers were sighted in two groups, one heading for the *Prince of Wales* and one for the *Repulse*. The *Prince of Wales* was slowed down, no longer under control. The Japanese were skimming the surface of the sea, and the list on the British ship made it impossible for her to depress her anti-aircraft guns suffi-ciently for them to hit the enemy, so that she was virtually without resistance when four torpedoes struck her at intervals along her hull. *Battleship* says: 'The whole effect of these four successive explosions on the starboard side of the ship was "quite stupendous, the ship appeared to jump sideways several inches, rather in the manner of an earth tremor".'

It has been estimated that, after receiving this broadside of torpedoes, the *Prince of Wales* had taken on board the total of 18,000 tons of water, while her speed had dropped to eight knots and she was settling deeper still into the water. Six Japanese planes had attacked, leaving twenty to deal with *Repulse*. Simultaneously these appeared on her port and starboard bows, so that when they dropped their torpedoes at close range it was impossible for *Repulse* to avoid them all and she was sunk at 1235. The *Prince of Wales* lasted nearly an hour longer and was still floating upside down when the fighters arrived from Singapore. A number of men had taken temporary refuge on the ship's bottom, where they could hear the shouts and cries of men trapped below, with no hope of rescue. The ship sank a few minutes later at approximately 1320.

Of the 1,612 men in the *Prince of Wales* 327 were lost, and of the 1,309 in the *Repulse* 513.

Appendix

Battleships sunk, wrecked or irreparably damaged, 1866–1955.
tb = torpedo-boat   d = destroyer   s/m = submarine   a/c = aircraft
cmb = coastal motor boat

					1800
1.	Re d'Italia	Italian	Rammed by enemy	Lissa	20/7/66
2.	Palestro	Italian	Gunfire	Lissa	20/7/66
3.	Captain	British	Capsized in gale	Biscay	7/9/70
4.	Vanguard	British	Collision	Irish Sea	1/9/75
5.	Grosser Kurfürst	German	Collision	Dover Strs	31/5/78
6.	Sultan	British	Wrecked (later salved)	Malta	6/8/89
7.	Blanco Encalada	Chilean	Torpedo (tb)	Caldera	23/4/91
8.	Victoria	British	Collision	Coast of Lebanon	22/6/93
9.	Aquidaban	Brazilian	Torpedo (tb)	Desterrol	14/4/94
10.	Ting Yuen	Chinese	Torpedo (tb)	Wei-hai-wei	5/2/95
11.	Maine	US	Internal explosion	Havana	15/2/98

					1900
12.	Petropavlovsk	Russian	Mine	Pt Arthur	13/4/04
13.	Hatsuse	Japanese	Mine	Pt Arthur	15/5/04
14.	Yashima	Japanese	Mine	Pt Arthur	15/5/04
15.	Poltava	Russian	Gunfire from shore	Pt Arthur	5/12/04
16.	Retvisan	Russian	Gunfire from shore	Pt Arthur	6/12/04
17.	Peresviet	Russian	Gunfire from shore	Pt Arthur	6/12/04
18.	Pobieda	Russian	Gunfire from shore	Pt Arthur	7/12/04

(Nos 15–18 were raised by the Japanese and incorporated in their navy. 15 and 17 were sold back to Russia in 1916 and 17 was lost in the following year. See No 55.)

19.	Sevastopol	Russian	Scuttled	Pt Arthur	1/1/05
20.	Kniaz Suvarov	Russian	Gunfire	Tsushima	27/5/05
21.	Osliabia	Russian	Gunfire	Tsushima	27/5/05
22.	Borodino	Russian	Gunfire	Tsushima	27/5/05
23.	Imperator Alexander III	Russian	Gunfire	Tsushima	27/5/05
24.	Sissoi Veliki	Russian	Gunfire	Tsushima	27/5/05
25.	Navarin	Russian	Gunfire	Tsushima	27/5/05
26.	Mikasa	Japanese	Internal explosion	Yokosuka	12/9/05
27.	Aquidaban	Brazilian	Internal explosion	(see No. 8)	21/1/06
28.	Montagu	British	Wrecked	Lundy	30/5/06
29.	Iéna	French	Internal explosion	Toulon	12/3/07

30. *Liberté*	French	Internal explosion	Toulon	25/9/11
31. *Audacious*	British	Mine	N. coast Ireland	25/10/14
32. *Bulwark*	British	Internal explosion	Sheerness	26/11/14
33. *Messudieh*	Turkish	Torpedo (s/m)	Dardanelles	13/12/14
34. *Formidable*	British	Torpedo (s/m)	English Channel	1/1/15
35. *Bouvet*	French	Mine	Dardanelles	18/3/15
36. *Irresistible*	British	Mine	Dardanelles	18/3/15
37. *Ocean*	British	Mine	Dardanelles	18/3/15
38. *Goliath*	British	Torpedo (d)	Dardanelles	13/5/15
39. *Triumph*	British	Torpedo (s/m)	Dardanelles	25/5/15
40. *Majestic*	British	Torpedo (s/m)	Dardanelles	27/5/15
41. *Khaireddin Barbarossa*	Turkish	Torpedo (s/m)	Dardanelles	8/8/15
42. *Benedetto Brin*	Italian	Sabotage	Brindisi	27/9/15
43. *King Edward VII*	British	Mine	N. of Scotland	16/1/16
44. *Russell*	British	Mine	Off Malta	27/4/16
45. *Indefatigable*	British	Gunfire	Jutland	31/5/16
46. *Queen Mary*	British	Gunfire	Jutland	31/5/16
47. *Invincible*	British	Gunfire	Jutland	31/5/16
48. *Lützow*	German	Gunfire	Jutland	1/6/16
49. *Pommern*	German	Torpedo (d)	Jutland	1/6/16
50. *Leonardo da Vinci*	Italian	Sabotage	Taranto	2/8/16
51. *Imperatritsa Maria*	Russian	Internal explosion	Sevastopol	20/10/16
52. *Suffren*	French	Torpedo (s/m)	N.W. Lisbon	26/11/16
53. *Regina Margherita*	Italian	Mine	Albanian coast	11/12/16
54. *Gaulois*	French	Torpedo (s/m)	Aegean	27/12/16
55. *Peresviet*	Russian	Mine	Off Pt Said (see No. 17)	4/1/17
56. *Cornwallis*	British	Torpedo (s/m)	Off Malta	9/1/17
57. *Tsukuba*	Japanese	Internal explosion	Yokosuka	14/3/17
58. *Danton*	French	Torpedo (s/m)	S. of Sardinia	19/3/17
59. *Vanguard*	British	Internal explosion	Scapa Flow	9/7/17
60. *Slava*	Russian	Gunfire	Moon Sound	17/10/17
61. *Wien*	Austrian	Torpedo (cmb)	Trieste	9–10/12/17
62. *Rheinland*	German	Stranded, salved but too badly damaged for further service	Aaland Is.	11/4/18
63. *Szent Istvan*	Austrian	Torpedo (cmb)	Adriatic	10/6/18
64. *Svobodnaya Rossia*	Russian	Torpedo (d) Sunk to prevent it passing under German control		18/6/18
65. *Kawachi*	Japanese	Internal explosion	Tokuyama Bay	12/7/18

66. *Viribus Unitis*	ex-Austrian On previous day had hoisted Yugoslav flag	Italian frogmen	Pola	1/11/18
67. *Britannia*	British	Torpedo (s/m)	Cape Trafalgar	9/11/18
68. *Mirabeau*	French	Stranded, salved but too badly damaged for further service	Black Sea	13/2/19
69. *Bayern*	German	Scuttled	Scapa Flow	21/6/19
70. *Prinzregent Luitpold*	German	Scuttled	Scapa Flow	21/6/19
71. *König Albert*	German	Scuttled	Scapa Flow	21/6/19
72. *Kaiser*	German	Scuttled	Scapa Flow	21/6/19
73. *Kaiserin*	German	Scuttled	Scapa Flow	21/6/19
74. *Friederich der Grosse*	German	Scuttled	Scapa Flow	21/6/19
75. *Markgraf*	German	Scuttled	Scapa Flow	21/6/19
76. *König*	German	Scuttled	Scapa Flow	21/6/19
77. *Grosser Kurfürst*	German	Scuttled	Scapa Flow	21/6/19
78. *Kronprinz Wilhelm*	German	Scuttled	Scapa Flow	21/6/19
79. *Hindenburg*	German	Scuttled	Scapa Flow	21/6/19
80. *Derfflinger*	German	Scuttled	Scapa Flow	21/6/19
81. *Seydlitz*	German	Scuttled	Scapa Flow	21/6/19
82. *Moltke*	German	Scuttled	Scapa Flow	21/6/19
83. *Von der Tann*	German	Scuttled	Scapa Flow	21/6/19
84. *Petropavlovsk*	Russian	Torpedo (cmb)	Kronstadt	17/8/19

(Raised and returned to service, sunk again World War II by bombs and shore artillery, 23/9/41.)

85. *Andrei Pervosvanni*	Russian	Torpedo (cmb)	Kronstadt	17/8/19
86. *Frunze*	Russian	Irreparably damaged by fire	Leningrad	1922
87. *France*	French	Wrecked	Quiberon Bay	26/8/22
88. *España*	Spanish	Wrecked	C. Tres Forcas (Morocco)	26/8/23
89. *España* (ex *Alfonso XIII*)	Spanish	Mine	Off Santander	30/4/37
90. *Jaime I*	Spanish	Internal explosion	Cartagena	17/6/37
91. *Royal Oak*	British	Torpedo (s/m)	Scapa Flow	14/10/39
92. *Norge*	Norw'gian	Torpedo (d)	Narvik	9/4/40
93. *Eidsvold*	Norw'gian	Torpedo (d)	Narvik	9/4/40
94. *Bretagne*	French	Gunfire	Oran	3/7/40
95. *Littorio*	Italian	Torpedo (a/c)	Taranto	12/11/40

96. *Duilio*	Italian	Torpedo (a/c) (Raised and returned to service)	Taranto	12/11/40
97. *Cavour*	Italian	Torpedo (a/c)	Taranto	12/11/40

(Raised and returned to service, sunk at Trieste by Italians to prevent her falling into German hands, raised by Germans and sunk a third time by bombers USAAF February 1945, raised a third time and broken up.)

98. *Kilkis*	Greek	Bombs (a/c)	Salamis	23/4/41
99. *Hood*	British	Gunfire	Off Iceland	24/5/41
100. *Bismarck*	German	Gunfire	N. Atlantic	27/5/41
101. *Barham*	British	Torpedo (s/m)	E. Mediterranean	25/11/41
102. *Arizona*	US	Torpedo (a/c)	Pearl Harbor	7/12/41
103. *Oklahoma*	US	Torpedo (a/c)	Pearl Harbor	7/12/41

The following five ships were all raised and returned to service:

104. *West Virginia*	US	Torpedo (a/c)	Pearl Harbor	7/12/41
105. *California*	US	Torpedo (a/c)	Pearl Harbor	7/12/41
106. *Nevada*	US	Torpedo (a/c)	Pearl Harbor	7/12/41
107. *Queen Elizabeth*	British	Italian frogmen	Alexandria	19/12/41
108. *Valiant*	British	Italian frogmen	Alexandria	19/12/41
109. *Hiei*	Japanese	Bombs (a/c)	Guadalcanal	12/11/42
110. *Kirishima*	Japanese	Gunfire	Guadalcanal	15/11/42
111. *Dunkerque*	French	Scuttled	Toulon	27/11/42
112. *Strasbourg*	French	Scuttled	Toulon	27/11/42
113. *Provence*	French	Scuttled	Toulon	27/11/42
114. *Mutsu*	Japanese	Internal explosion	Hiroshima Bay	8/6/43
115. *Roma*	Italian	Glider bomb	Off Corsica	9/9/43
116. *Scharnhorst*	German	Gunfire	North Cape	26/12/43
117. *Courbet*	French	Scuttled as part Mulberry 'A'	Arromanches	10/6/44
118. *Musashi*	Japanese	Bombs and Torpedoes (a/c)	Leyte	24/10/44
119. *Fuso*	Japanese	Torpedo (d)	Leyte	25/10/44
120. *Yamashiro*	Japanese	Torpedo (d) and Gunfire	Leyte	25/10/44
121. *Tirpitz*	German	Bombs	Tromso	12/11/44
122. *Kongo*	Japanese	Torpedo (s/m)	Formosa	21/11/44
123. *Schleswig Holstein*	German	Bombs	Gdynia	18/12/44
124. *Gneisenau*	German	Scuttled	Gdynia	4–5/4/45
125. *Yamato*	Japanese	Torpedo and Bombs (a/c)	Okinawa	7/4/45
126. *Schlesien*	German	Bombs	Swinemünde	4/5/45
127. *Hiuga*	Japanese	Bombs	Kure	24/7/45
128. *Ise*	Japanese	Bombs	Kure	28/7/45
129. *Haruna*	Japanese	Bombs (a/c)	Kure	28/7/45
130. *Novorossisk* (ex *Cesare*)	Russian	Internal explosion	Sevastopol	29/10/55

Notes

1. The first and subsequent days of the landings were referred to as A day, A + 1, A + 2 etc., by decision of General MacArthur who did not wish his operations to be confused with those of General Eisenhower in Europe.
2. C. Vann Woodward, *The Battle for Leyte Gulf* (New York, Macmillan, 1947), p. 235.
3. Theodore Roscoe, *US Submarine Operations in World War II* (Annapolis, Md., US Naval Institute, 1949).
4. 4,950 tons with four 9-inch guns.
5. G. E. Armstrong, *Torpedoes and Torpedo Vessels* (London, Bell, 1901).
6. A cofferdam is a box-like structure which can be fitted over the damaged side of a ship below the water-line, and then pumped out so that repairs can be carried out.
7. Captain Damon E. Cummings, USN (retd), *Admiral Richard Wainwright and the United States Fleet* (Washington DC, US Government Printing Office, 1962).
8. Admiral H. G. Rickover, USN, *How the Battleship 'Maine' was destroyed* (Washington DC, Naval History Division, Dept. of the Navy, 1976), p. 106.
9. France honoured this officer in 1915 by giving his name to a new destroyer.
10. The commanding officer of the *Liberté*, on leave at the time of the disaster, was Commandant Jaurès, who came of an old naval family and was the brother of the famous Socialist politician.
11. Dreuzy adds, 'I had this from Garnier himself.'
12. H. H. Herwig, *Luxury Fleet* (London, George Allen & Unwin, 1980).
13. Immediately after the end of the war there was a wholesale change of names by many of the principal towns and areas on the shores of the Baltic. Reval became Tallin, Helsingfors became Helsinki, Libau became Liepaja, Memel became Klaipeda etc. After World War II there was another rash of changes, so that if the reader is using a modern (i.e. post-1945) atlas he will find Gdansk for Danzig, Kaliningrad for Königsberg, Szczecin for Stettin and many more.
14. Her Commanding Officer was Captain E. G. Kennedy who lost his life when in command of the armed merchant-cruiser *Rawalpindi*, sunk by the German battle-cruisers *Scharnhorst* and *Gneisenau* in November 1939.
15. Captain S. W. Roskill, DSC, RN, *The War at Sea, 1939–45: Vol. I The Defensive* (London, HMSO, 1977).
16. Ibid, pp. 397–8.
17. Double British Summer Time, then being used in the British ships, was two hours ahead of Greenwich Mean Time. The result of this was that, by ships' time, sunset had been at 1.51 am on 24 May and sunrise at 6.37 am.
18. *The Encyclopaedia of World War II* (Secker & Warburg, 1978), p. 183.

Index